Math Mammoth
Grade 5-B Worktext

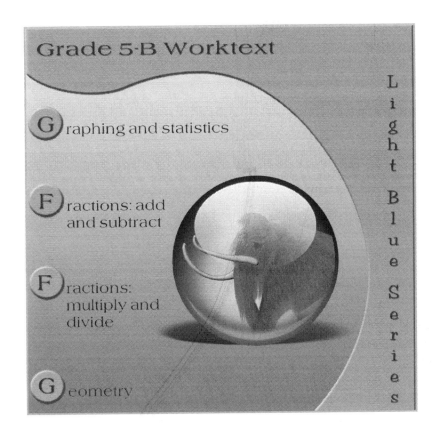

Grade 5·B Worktext

G raphing and statistics

F ractions: add and subtract

F ractions: multiply and divide

G eometry

Light Blue Series

By Maria Miller

Contents

Chapter 7: Fractions: Multiply and Divide

Chapter 8: Geometry

Foreword

Math Mammoth Grade 5-B Worktext comprises a complete math curriculum for the second half of fifth grade mathematics. Fifth grade focuses on fractions and decimals, in particular. In part 5-A, students have studied the four operations with whole numbers, large numbers, problem solving, and decimal arithmetic. In this part, 5-B, we study common statistical graphs, fraction arithmetic, and geometry.

This book starts with chapter 5, where we study graphing in a coordinate grid, line and bar graphs, and average and mode. Today's world has become increasingly complex with lots of data presented in the media, so our children need a good grasp of statistical graphs to be able to make sense of all of that information.

Chapter 6 covers the addition and subtraction of fractions—another topic of focus for 5th grade, besides decimals. The most difficult topic of this chapter is adding and subtracting unlike fractions, which is done by first converting them to equivalent fractions with a common denominator.

In chapter 7, we study the multiplication and division of fractions (division only in special cases), relying first on visual models, and then proceeding to the abstract shortcuts.

Chapter 8 takes us to geometry, starting with a review of angles and polygons. From there, students will learn to draw circles, to classify triangles and quadrilaterals, and the concept of volume in the context of right rectangular prisms (boxes).

I wish you success with teaching math!

Maria Miller, the author

Chapter 5: Statistics and Graphing
Introduction

This chapter starts out with a study of the coordinate grid, but only in the first quadrant. Besides learning how to plot points, students also plot ordered pairs (points) from number patterns or rules. This is actually the beginning of the study of *functions*.

Practicing the use of the coordinate grid is a natural "prelude" to the study of line graphs, which follows next. The goals are that the student will be able to:

- read line graphs, including double line graphs, and answer questions about data already plotted;

- draw line graphs from a given set of data.

The goals for the study of bar graphs are similar to those for the study of line graphs, in that the student will need to both:

- read bar graphs, including double bar graphs, and answer questions about data already plotted; and

- draw bar graphs and histograms from a given set of data.

In order to make histograms, it is necessary to understand how to group the data into categories ("bins"). The lesson *Making Histograms* explains the method we use to make categories if the numerical data is not already categorized.

Toward the end of the chapter, we study average (also called the *mean*) and mode, and how these two concepts relate to line and bar graphs. Other math curricula commonly introduce the median, too, but I decided to omit it from 5th grade. There is plenty of time to learn that concept in subsequent grades. Introducing all three concepts at the same time tends to jumble the concepts together and confuse them —and all a lot of students are able to grasp from that is only the calculation procedures. I feel it is better to introduce and contrast initially only the two concepts, the mean and the mode, in order to give the student a solid foundation. We will introduce the median later, and then compare and contrast it with the other two.

This chapter also includes an optional statistics project, in which the student can develop investigative skills.

The Lessons in Chapter 5

Helpful Resources on the Internet

COORDINATE GRID

Billy Bug Game
Click on the arrow keys to guide Billy to the coordinates of the hidden grub. How long will it take you to feed Billy ten times?
https://www.math10.com/en/math-games/games/geometry/games-billy-the-bug.html

Soccer Coordinates Game
Plot the coordinates on the coordinate grid correctly to block the soccer ball from entering the goal.
http://www.xpmath.com/forums/arcade.php?do=play&gameid=90

Coordinate Grid Quiz from ThatQuiz.org
Practice plotting a point and giving the coordinates of a given point (in the first quadrant).
https://www.thatquiz.org/tq-7/?-j48-l5-p0

Function Machine
Enter a rule in the function machine. Then enter various x-values to find their corresponding y-values, and generate a table and a graph of the function.
http://hotmath.com/util/hm_flash_movie.html?movie=/learning_activities/interactivities/function_machine.swf

Function Machine
What's the rule? Enter your own values or let the computer decide for you.
http://www.mathplayground.com/functionmachine.html

Number Pattern Tables
Apply the rule to find the missing number in the table.
https://www.studyladder.com/games/activity/number-pattern-tables--20584

Interpret Relationships Between Number Patterns
Generate patterns using given rules, identify relationships between terms, and graph ordered pairs consisting of corresponding terms from the patterns.
https://www.khanacademy.org/math/pre-algebra/applying-math-reasoning-topic/number-patterns/e/visualizing-and-interpreting-relationships-between-patterns

Graph a Two-Variable Relationship
Practice identifying relationships between variables with this interactive exercise.
https://www.ixl.com/math/grade-5/graph-a-two-variable-relationship

GRAPHING AND GRAPHS

Easy Practice Problems for Reading Bar Graphs
First, customize your bar chart. Then, click on the buttons on the left side to get questions to answer.
http://www.topmarks.co.uk/Flash.aspx?f=barchartv2

Graphs Quiz from That Quiz.org
Questions about different kinds of graphs (bar, line, circle graph, multi-bar, stem-and-leaf, box plot, scatter graph). You can modify the quiz parameters to your liking.
http://www.thatquiz.org/tq-5/math/graphs

Survey Game
First, ask children their favorite hobby or color. Then, make a frequency table, a bar graph, and a pictogram from the results.
http://www.kidsmathgamesonline.com/numbers/mathdata.html

Line Graphs Quiz
Answer the questions about the line graph in this interactive 10-question quiz.
https://www.thatquiz.org/tq-5/?-j10f14-l5-p0

Creating Histograms at Khan Academy
Use the given data to create a histogram in this interactive exercise.
https://www.khanacademy.org/math/cc-sixth-grade-math/cc-6th-data-statistics/histograms/e/creating-histograms

Create Double Bar Graphs Using Tables
Use the data in the table to complete the bar graph in this interactive exercise.
http://www.mathgames.com/skill/6.96-create-double-bar-graphs-using-tables

Interpret Double Bar Graphs
Read the double bar graphs and answer the questions in this interactive quiz.
http://www.mathgames.com/skill/6.95-interpret-double-bar-graphs

Statistics Interactive Activities from Shodor
A set of interactive tools for exploring and creating different kinds of graphs and plots. You can enter your own data or explore the examples.

http://www.shodor.org/interactivate/activities/BarGraph/

http://www.shodor.org/interactivate/activities/Histogram/

http://www.shodor.org/interactivate/activities/CircleGraph/

http://www.shodor.org/interactivate/activities/MultiBarGraph/

Math Goodies Interactive Data and Graphs Lessons
Clear lessons with examples and interactive quiz questions, covering the concept and construction of line graphs, bar graphs, circle graphs, comparing graphs, and exercises.
http://www.mathgoodies.com/lessons/toc_vol11.html

Interactive tool for creating graphs
Customize your own bar graph, line graph, or pie chart using this interactive tool.
https://www.mathsisfun.com/data/data-graph.php

Create a Graph
Create bar graphs, line graphs, pie graphs, area graphs, and xyz graphs to view, print, and save.
http://nces.ed.gov/nceskids/createagraph/default.aspx

Data Grapher
Use this tool to create bar graphs, line graphs, pie charts, and pictographs. You can enter multiple rows and columns of data, select which set(s) to display in a graph, and choose the type of representation.
http://illuminations.nctm.org/Activity.aspx?id=4098

MEAN, MEDIAN, MODE, AND RANGE

Math - Elephants - Line Graphs & Mean
Interactive exercises for interpreting a line graph, drawing a line graph, and calculating the mean.
http://www.e-learningforkids.org/math/lesson/elephants-plant-line-graphs-mean/

Mean/Mode Quiz
A 10-question quiz about calculating the mode and the mean.
http://www.thatquiz.org/tq-p-z1/?-j6g00-l5-p0

The Mean Machine
Use this interactive tool to see how average is calculated.
http://www.mathsisfun.com/data/mean-machine.html

Quiz: Finding the Mean of a Set of Numbers
Practice calculating the mean for a simple data set. This quiz helps to clarify the definition of mean as it relates to median, mode and range.
http://www.turtlediary.com/game/finding-the-mean-of-set-of-numbers.html

Study Jams: Mode
This site gives step-by-step illustrations of how to find the mode for a set of data.
http://studyjams.scholastic.com/studyjams/jams/math/data-analysis/mode.htm

Mean, Median, and Mode
How to calculate the mean, the median, and the mode for sets of data given in different ways. There are also interactive exercises.
http://www.cimt.org.uk/projects/mepres/book8/bk8i5/bk8_5i2.htm

Using and Handling Data
Simple explanations for finding the mean, median, or mode.
http://www.mathsisfun.com/data/central-measures.html

Measures Activity
Enter your own data and the program will calculate the mean, median, mode, range, and some other statistical measures.
http://www.shodor.org/interactivate/activities/Measures

Coordinate Grid

This is a **coordinate grid**.

The long black line across the bottom, with the "*x*" near its end, is called the ***x*-axis**.

The vertical line that has "*y*" near the top is called the ***y*-axis**.

You can see one point, called "A," that is drawn or *plotted* on the grid.

It has two numbers *associated*, or matched, with it. Those two numbers are called the **coordinates** of the point A.

The first number is called the **x-coordinate** of the point A, and the second number is called the **y-coordinate** of the point A.

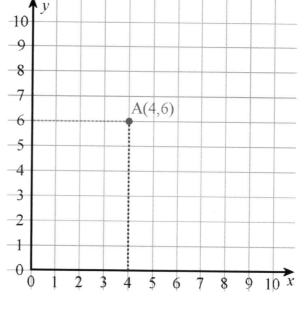

The x-coordinate of the point A is 4 because if you drew a line straight down from A, it would *intersect*, or "hit," the x-axis at 4. The y-coordinate of the point A is 6 because if you drew a line straight left from A, it would intersect the *y*-axis at 6.

We write the two coordinates of a point inside parentheses, separated by a comma.

Note: The order of the two coordinates matters. The *first* number is ALWAYS the *x*-coordinate, and the *second* number is ALWAYS the *y*-coordinate, not the other way around.
So (5, 8) means that the x-coordinate is 5 and the y-coordinate is 8.

1. Write the two coordinates of the points plotted on the coordinate grid. For points A and B, the helping lines are drawn in.

A (___ , ___)   B (___ , ___)

C (___ , ___)   D (___ , ___)

E (___ , ___)   F (___ , ___)

G (___ , ___)   H (___ , ___)

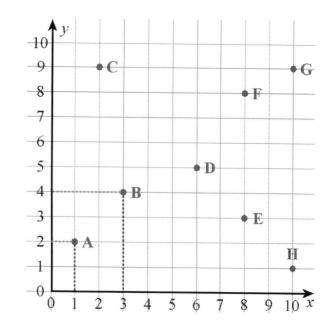

Notice especially the points that are located on the two axes.

If a point lies on the *y*-axis, its *x*-coordinate is zero. A is (0, 6) and B is (0, 3) .

If the point lies on the *x*-axis, its *y*-coordinate is zero. D is (5, 0) and E is (9, 0).

The point C has the coordinates (0, 0). This point (0, 0) is called the **origin**.

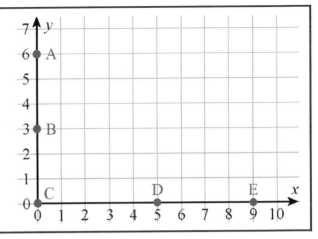

2. Plot and label the following points on the coordinate grid.

A (2, 8)   B (0, 5)   C (4, 0)

D (9, 10)   E (8, 5)   F (1, 4)

G (1, 0)   H (0, 8)   I (3, 7)

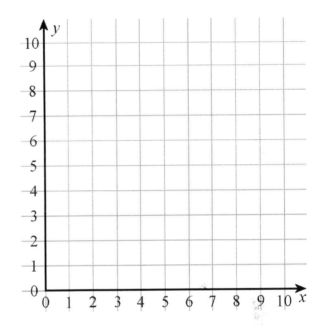

3. The coordinate grid is very useful for many things. For example, computer drawing programs use it frequently. Let's say "LINE (5,6) - (2,7)" means a straight line segment that is drawn from the point (5, 6) to the point (2, 7).

Draw the following line segments.
What figure is formed?

LINE (1, 0) - (7, 0)   LINE (7, 0) - (7, 5)

LINE (1, 0) - (1, 5)   LINE (1, 5) - (0, 5)

LINE (0, 5) - (4, 7)   LINE (4, 7) - (8, 5)

LINE (8, 5) - (7, 5)   LINE (3, 0) - (3, 3)

LINE (5, 0) - (5, 3)   LINE (3, 3) - (5, 3)

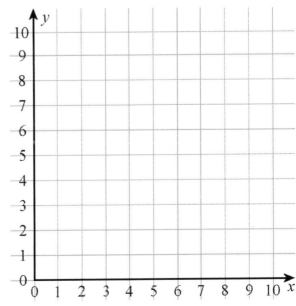

This example shows point A moving four units down and then two units to the right. The new location is called point A′ (read "A prime").

Originally A's coordinates were (1, 6).
After the movement, the coordinates are (3, 2)

Notice how you can just subtract four units from the y-coordinate (the movement four units straight down) and add two units to the x-coordinate (movement two units to the right).

Point B is originally at (5, 7). It moves four units to the right and two up. You add four to the x-coordinate, and two to the y-coordinate. Its new coordinates are (9, 9).

Movement up or down affects the y-coordinate.
Movement right or left affects the x-coordinate.
In other words, movement *parallel* to an axis affects that same coordinate.

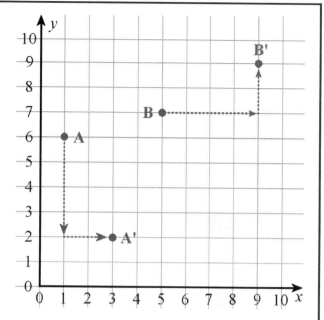

4. The three vertices of a triangle are (2, 0), (5, 1) and (3, 4). The triangle is moved three units to the right and two up.

   a. Plot the vertices of the triangle before and after the movement.

   b. Write the coordinates of the vertices after the movement.

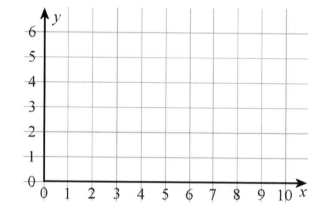

5. **a.** Determine how the line segment has been moved, and move the triangle ABC the same way. Let's call the new triangle A′B′C′. Write the coordinates of the vertices of the triangle A′B′C′ after the movement.

   b. Let's say the point (3, 5) moves to (2, 7). Move the triangle ABC in a similar way. Write the coordinates of the triangle's vertices after the movement.

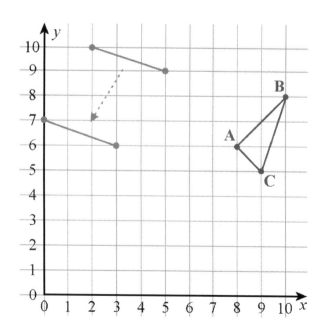

12

Number Patterns in the Coordinate Grid

Remember the "number rules" we studied a little while ago? There is something special about those rules and the coordinate grid. In the number rules lesson, the two numbers were labeled A and B. This time we label them x and y so that we get **number pairs**, and we can then **plot** those on the coordinate grid.

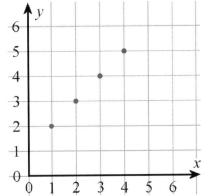

x	1	2	3	4
y	2	3	4	5

The rule is: y is 1 more than x, or $y = x + 1$.

From the table we get lots of number pairs. Just take each x and pair it with its corresponding y to get the number pairs (1, 2), (2, 3), (3, 4), and (4, 5). Those four number pairs are <u>four points</u> on the coordinate grid (see the image).

1. Plot the points from the "number rules" or number patterns on the coordinate grids.

a.

x	0	1	2	3	4	5	6
y	3	4	5	6	7	8	9

The rule is: $y = x + 3$.

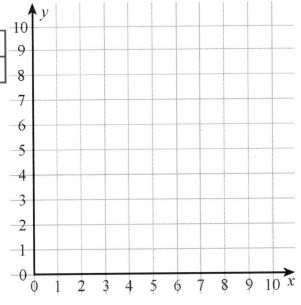

b.

x	0	1	2	3	4	5	6
y	6	5	4	3	2	1	0

The rule is: x and y always add up to 6.
In other words, $x + y = 6$.

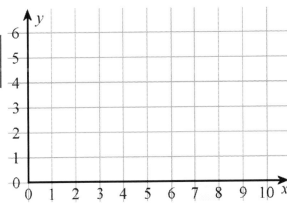

2. Plot the points from the number rules on the coordinate grids.
   Fill in the rest of the table first, using the rule.

a.

x	0	2	4	6	8	10
y	0	1	2			

The rule is: y is half x. In another way, $x = 2y$.

> **Note:** The expression "2y" means "2 times y" ("2 × y"). The multiplication sign is left off between a number and a letter.

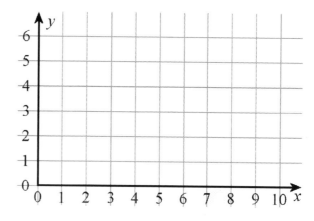

b.

x	2	3	4	5	6	7
y	10	8	6			

The rule is: $y = 14 - 2x$. Or, to get y, you double x and subtract that from 14.

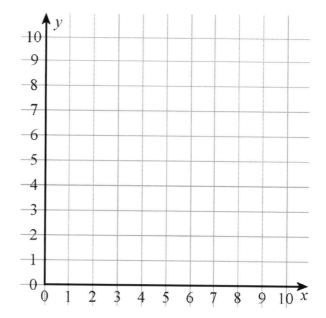

c.

x	1	2	3	4	5	6	7	8	9	10
y	4			3			9			

The rule is: Choose y randomly from the whole numbers 0, 1, 2, 3, 4, 5, 6, 7, 8, 9 and 10. In other words, let y be any whole number you like between 0 and 10. (It can be different each time, or it can be same.)

That is a really funny "rule," isn't it?

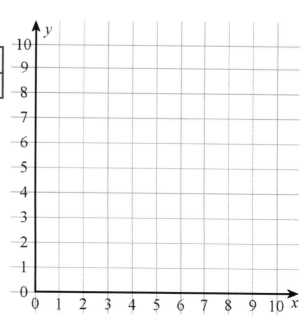

14

3. This time the coordinate grid is *scaled* differently.
   The x-values and the y-values written next to the axes do not go by ones. We may have to place some points in between the gridlines. See for example the point (10, 3).

   Fill in the number table and plot the points.

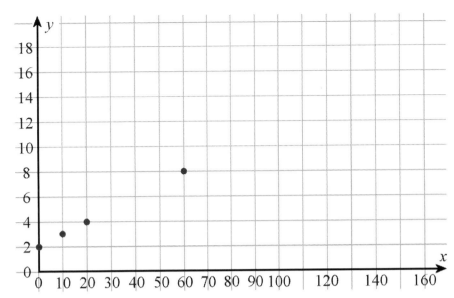

x	0	10	20	30	40	50	60	70	80	90	100	110
y	2	3	4				8					

The rule is: $y = \dfrac{x}{10} + 2$. (First divide x by 10, then add 2.)

4. Write the number pairs in the table, using the plot. Then, write the "number rule."

   a.

x						
y						

   The rule is:

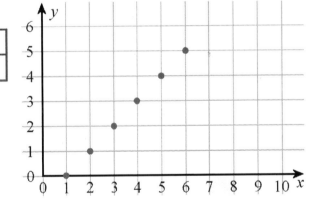

   b.

x					
y					

x					
y					

   The rule is:

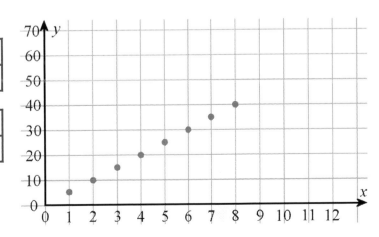

15

These tables with *x*'s and *y*'s are *functions*. All through this lesson you have been plotting functions.

A **function** is simply a collection of number pairs—with one special limitation. And that is that you cannot have two number pairs with same *x*'s but different *y*'s. For example, you cannot have both (2, 4) and (2, 1) in the same function.

Functions (collections of number pairs) can be presented in various ways. One way is the "number rule table." Another way is just writing the number pairs as a list: (6, 7), (5, 4), and (3, 1) is a function.

Yet another way is by specifying what the first numbers in the number pairs should be (the *x*'s) and giving a rule for the relationship between each *x* and *y*. For example, this is a function: Let $y = x + 8$, and *x* is all the whole numbers from 15 to 25. (Can you figure out the number pairs this function has?)

Yet one more way to present a function is to plot the number pairs as points on the coordinate grid. The plot on the right gives us the function (2, 2), (3, 4), (4, 5), (5, 4), and (6, 2).

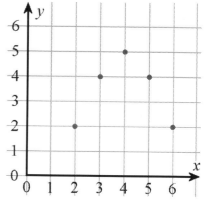

5. The plot on the right defines a certain function.

   a. Give this function as a list of number pairs.

   b. Give this function in a table. Also write a number rule for it.

x						
y						

   The rule is: _____

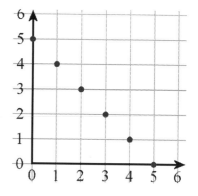

6. Make your own function. Call it "MyFunction". Represent MyFunction in three different ways:

   a. As a list of number pairs:

   b. As a table:

   c. As a plot.

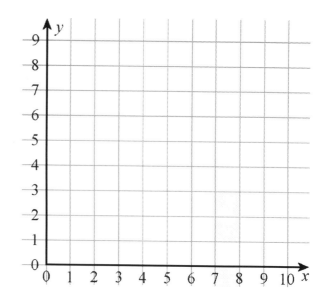

16

More Number Patterns in the Coordinate Grid

This time we will make a "number rule" in a little different way. We will first make a list of *x*-values using some pattern or rule. Then we will do the same for *y*-values.

Example 1.
The rule for *x*-values: start at 0, and add 2 each time.
The rule for *y*-values: start at 10, and subtract 1 each time.

x	0	2	4	6	8	10
y	10	9	8	7	6	5

We plot the number pairs. Notice, they form as if it were a line.

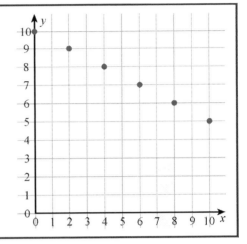

Plot the points from the "number rules" or number patterns on the coordinate grids.

1. The rule for *x*-values: start at 0, and add 1 each time.
   The rule for *y*-values: start at 1, and add 2 each time.

x	0	1	2			
y	1	3				

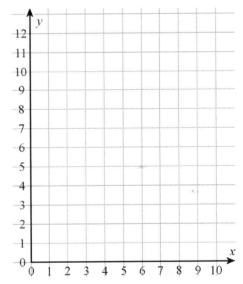

2. The rule for *x*-values: start at 10, and subtract 1 each time.
   The rule for *y*-values: start at 1, and add 2 each time.

x						
y						

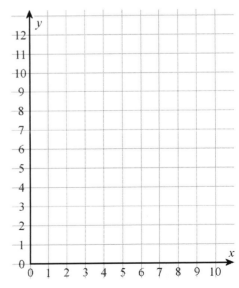

17

Plot the points from the number rules on the coordinate grids.

3. <u>The rule for *x*-values</u>: start at 1, and add 1 each time.
   <u>The rule for *y*-values</u>: start at 5, and subtract ½ each time.

x								
y								

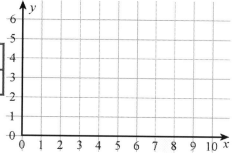

4. <u>The rule for *x*-values</u>: start at 8, and subtract ½ each time.
   <u>The rule for *y*-values</u>: start at 0, and add 1 each time.

x								
y								

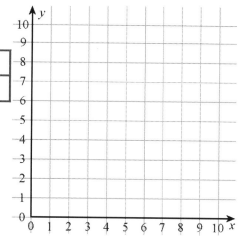

5. Make your own rule.

   <u>The rule for *x*-values</u>: start at _____, and

   <u>The rule for *y*-values</u>: start at _____, and

x					
y					

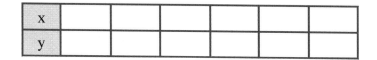

6. Make your own rule. Plot the points in the same
   grid as above or in the small grid (if they fit).

   <u>The rule for *x*-values</u>: start at _____, and

   <u>The rule for *y*-values</u>: start at _____, and

x					
y					

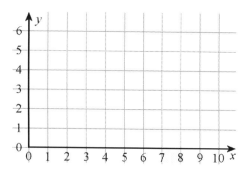

Example 2.

The rule for *x*-values:
start at 0, and add 3 each time.

The rule for *y*-values:
start at 0, and add 1 each time.

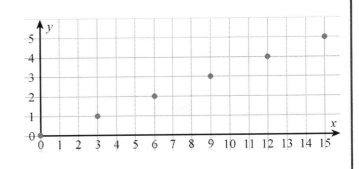

x	0	3	6	9	12	15
y	0	1	2	3	4	5

Notice that in each case, the *y*-coordinate is 1/3 of the *x*-coordinate! Or, the *x*-coordinate is three times the *y*-coordinate. We can write this as an equation: $y = \dfrac{x}{3}$ or $x = 3y$. Why is that?

Because if you add 3 each time, you get the skip-counting pattern by 3s, and if you add 1 each time, you are just counting by ones. Skip-counting by 3 makes the multiplication table of 3, so it makes sense that the rule tying *x* and *y* together has to do with multiplying or dividing by 3.

Plot the points from the number rules on the coordinate grids.

7. The rule for *x*-values: start at 0, and add 2 each time.
   The rule for *y*-values: start at 0, and add 1 each time.

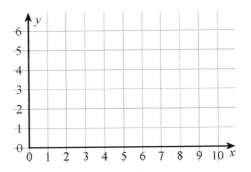

x						
y						

What simple rule ties the *x* and *y*-coordinates together in each case? In other words, how would you get *y* from *x*?

Explain in your own words why this is so.

8. The rule for x-values: start at 0, and add ½ each time.
   The rule for y-values: start at 0, and add 1 each time.

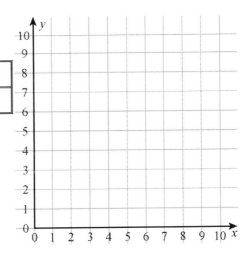

x								
y								

What simple rule ties the *x* and *y*-coordinates together in each case? In other words, how would you get *y*, knowing *x*?

Explain in your own words why this is so.

You can make more of your own number rules and plot them using grid paper (graph paper).

9. This time the coordinate grid is *scaled* differently.

The rule for *x*-values:
start at 0, and add 10 each time.

The rule for *y*-values:
start at 0, and add 1 each time.

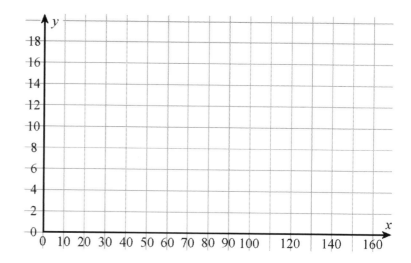

x	0	10	20	30								
y												

What simple rule ties the *x* and *y*-coordinates together in each case?
In other words, how would you get *y*, knowing *x*?

Explain in your own words why this is so.

10. One more!

The rule for *x*-values:
start at 0, and add 2 each time.

The rule for *y*-values:
start at 0, and add ½ each time.

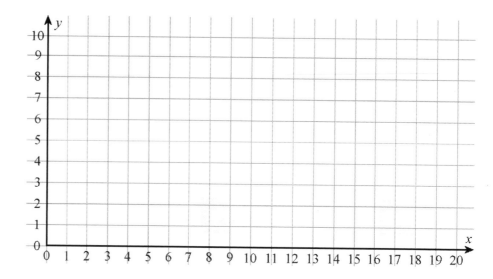

x											
y											

What simple rule ties the *x* and *y*-coordinates together in each case?
In other words, how would you get *y*, knowing *x*?

Explain in your own words why this is so.

Line Graphs

Mary sold muffins every day at 2 pm in the school cafeteria. She recorded her sales in the table.

Muffin Sales, Week 11	
Day	Muffins sold
Mon	24
Tue	36
Wed	41
Thu	33
Fri	17

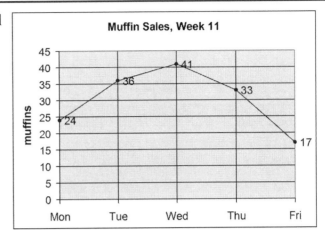

We can draw a line graph out of this data because the data is organized by *time* (days of the week). To do that, we first plot the individual data points in a coordinate grid. Then we draw lines to connect neighboring points.

Besides that, the line graph also needs:

- **a title** on top of the whole graph
- **labels** for the **tick marks** on the two axes
- a **label** for the **vertical axis** (the *y*-axis)
- a **label** for the **horizontal axis** (the *x*-axis) unless it is very clear what it is about. In the graph above, the labels "Mon," "Tue," and so on show very clearly that they are days of the week, so we don't necessarily need a title "Days of the week" for the horizontal axis.

Use a line graph for data that is organized by some unit of time (hours, days, weeks, years, *etc.*)

1. **a.** Add a label for the vertical axis that says "Rainfall (mm)". The "mm" stands for millimeters.

   b. Add five more data points to the graph according to this data:

Day	11	12	13	14	15
rainfall (mm)	9	0	0	13	2

   c. Draw a line between each two consecutive points.

   d. How many dry days were there in the first half of April?

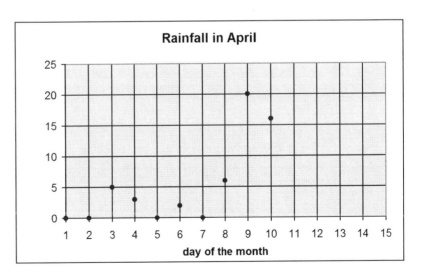

2. Jessie recorded the temperature of his fridge every 30 minutes during the day.
   a. Finish drawing the line graph.

   b. How did the temperature change around noon?
   Give a possible reason for that change.

   c. How did the temperature change around 5PM?
   Give a possible reason for that change.

Time	7:00	7:30	8:00	8:30	9:00	9:30	10:00	10:30	11:00	11:30	12:00	12:30
Temperature (°F)	46	47	52	50	49	47	50	50	48	56	62	55

Time	13:00	13:30	14:00	14:30	15:00	15:30	16:00	16:30	17:00	17:30	18:00	18:30	19:00
Temperature (°F)	50	48	47	51	50	48	53	55	64	61	55	50	46

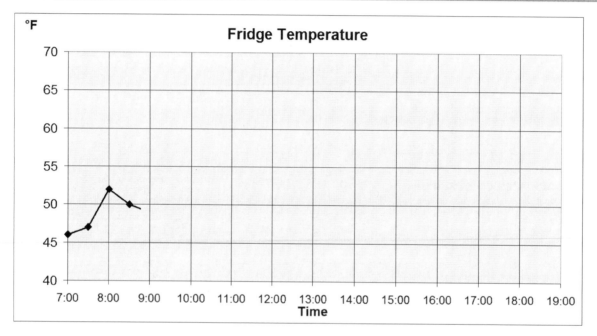

3. Robert recorded his total savings at the end of each month. Draw a line graph of that data.
   <u>Note:</u> *You* need to choose the scaling for the vertical axis so that the largest number, $107, will fit on the grid. Think: should the gridlines go by five? By ten? By fifteen? By some other number?

Month	Total savings
Apr	$8
May	$22
Jun	$46
Jul	$61
Aug	$78
Sep	$95
Oct	$107

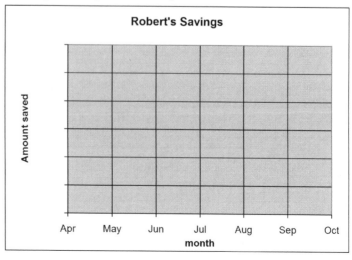

22

4. The table below shows the monthly visitor count to Juanita's blog. Three of the data points are already plotted. Your task is to plot the rest and finish the line graph.

Note: Since the vertical gridlines go by 200s, you cannot make an exact dot at, say, 1442. You need to round the numbers first. Round them to the nearest 50. Then plot the points.

Month	Visitors	rounded to the nearest 50
Jan	1039	1050
Feb	1230	1250
Mar	1442	
Apr	1427	
May	1183	1200
Jun	823	
Jul	674	650
Aug	924	
Sep	1459	
Oct	1540	
Nov	1638	
Dec	1149	

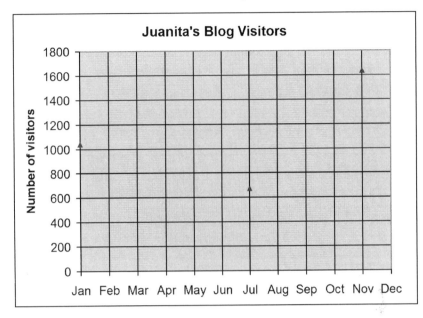

In the _____ Juanita's blog had far fewer visitors than in the spring or fall.

The three months with the fewest visitors were _____, _____, and _____.

The three months with the most visitors were _____, _____, and _____.

5. A car travels at a constant speed of 30 meters in each second.

Time	Distance
0 s	0 m
1 s	30 m
2 s	60 m
3 s	90 m
4 s	120 m
5 s	150 m

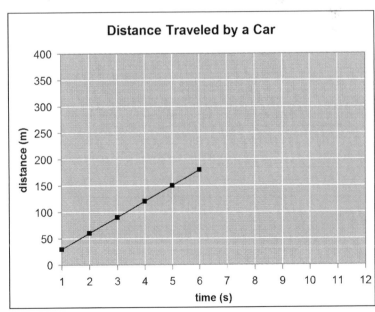

a. Fill in the table, and continue the graph till 12 seconds.

b. When will the car have traveled 3 km?

6. **a.** Draw a line graph of the data on the right.

After-School Sports Club	
Year	**Members**
1998	56
1999	63
2000	60
2001	35
2002	27
2003	32
2004	57
2005	63

- First draw the two axes, one at the bottom and the other at the left side. Use a ruler so the graph looks neat and tidy.

- Label the axes. Label the horizontal axis as "year" (not as "*x*"). Label the vertical axis as "members" (not as "*y*").

- Label the whole graph by writing at the top: "After-School Sports Club Members from 1998 to 2005."

- Since the horizontal axis is for the years, draw tick marks on that axis for the years, but use *three* squares between each tick mark because the numbers for the years are so long (four digits).

- Then choose a scaling for the vertical axis. Because the member counts vary from 27 to 63, it makes sense to mark the vertical axis in fives, starting from 0. In other words, let each grid square be 5 members.

- Now you are ready to plot the points and draw the line graph.

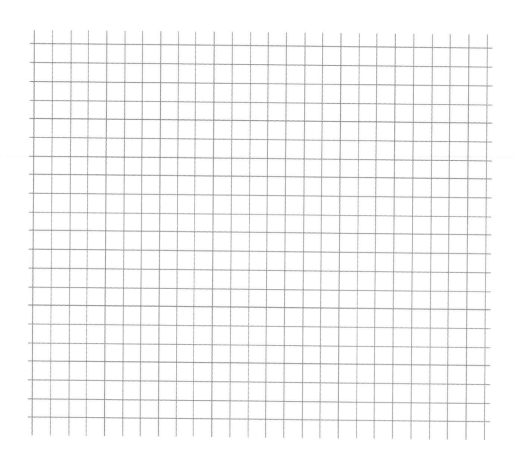

b. What do you think might have caused the drop in membership in 2001 - 2003?

Reading Line Graphs

The graph shows how many people were living on farms in the United States during 1900-2000. You can see how dramatically the number has dropped!

The question (a) in exercise 1 asks you to *estimate* the farm population in year 2010. Do it by tracing over the graph and continuing the graph in a natural way till the year 2010. The plain numbers listed in the table do not really help with estimation (without further mathematical tools).

Notice that the table lists the farm population in *thousands of people*. For example, in year 1970 there were 9712 thousand people—or 9,712,000 people—living on farms. In other words, you need to tag three zeros onto each of those numbers to get the numbers in reality.

Note also that these numbers are actually rounded to the nearest thousand—no population is an exact number of so many thousand people, year after year.

Year	Farm Population (thousands of people)
1900	29875
1910	32077
1920	31974
1930	30529
1940	30547
1950	23048
1960	13445
1970	9712
1980	6051
1990	4591
2000	2993

Source: Census of Agriculture

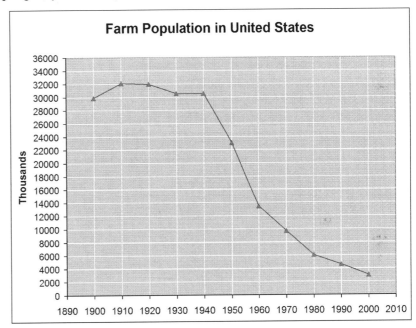

Farm Population in United States

1. **a.** Consider the graph above. Estimate the U.S. farm population in the year 2010.

   b. In which two decades were the greatest drops in farm population?

   c. By how many people did the farm population decrease during each of those two decades?

   d. What was the first year when the farm population dropped below 10 million?

   e. When approximately did the farm population drop below 5,000,000?

2. The International Union for Conservation of Nature (IUCN) produces a report every few years called *IUCN Red List of Threatened Species*. This report lists the number of animal and plant species that are considered endangered and extinct. The term "threatened" actually means the species can either be considered "Critically Endangered," "Endangered," or "Vulnerable."

Study the data and the graph below, and answer the questions.

Numbers of threatened species by major groups of organisms (1996–2012)

Number of threatened species →	in 1996/98	in 2000	in 2002	in 2003	in 2004	in 2006	in 2007	in 2008	in 2009	in 2010	in 2011	in 2012
Mammals	1,096	1,130	1,137	1,130	1,101	1,093	1,094	1,141	1,142	1,131	1,134	1,139
Birds	1,107	1,183	1,192	1,194	1,213	1,206	1,217	1,222	1,223	1,240	1,240	1,313
Reptiles	253	296	293	293	304	341	422	423	469	594	664	807
Fishes	734	752	742	750	800	1,171	1,201	1,275	1,414	1,851	2,011	2,058

(Data from 2012 IUCN Red List)

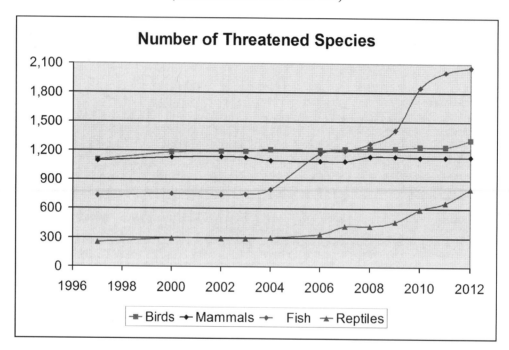

a. How many reptile species were considered threatened in 2003? In 2012?

b. How many fish species were considered threatened in 2003? In 2012?

c. In which major animal group has the number of threatened species stayed approximately the same between 2000 and 2012?

d. In which major animal groups has the number of threatened species nearly tripled from 2000 to 2012?

e. Find a period of time when the number of threatened species decreased in a certain species for several years. Which species was it, and when was it?

Double and Triple Line Graphs

A double-line graph shows data for two different things that occur in the same time period. It simply has one line for one set of data, and another line for the other.

We can distinguish these two lines by marking the data points in different manners and by using different colors. For example, we can use circles for one data set and triangles for the other.

Usually double-line graphs also have a *legend* that explains which line belongs to which data set.

The same principles apply to triple-line graphs or quadruple line graphs.

In the example here, we can see that Mom sent many more messages than Dad.

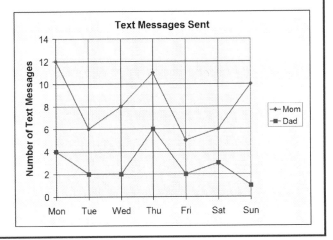

1. Refer to the double-line graph above about the text messages Mom and Dad sent with their cell phones.

   a. Find how many messages Mom sent in total, and how many Dad sent in total.

   b. How many more messages did Mom send on Thursday than Dad?

   c. Find the day with the *greatest difference* between the number of messages Mom sent and the number of messages Dad sent.

   d. Find the day with the *least difference* between the number of messages Mom sent and the number of messages Dad sent.

2. This graph shows the total number of tropical storms and hurricanes in three Atlantic hurricane seasons.

   a. Find the total number of storms for 2005, 2006, and 2007.

   b. Which year was unusually active?

   c. Based on this graph, which month is the most active month?

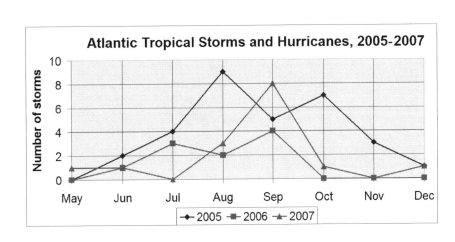

3. Project. (optional)
   Go to http://en.wikipedia.org/wiki/Timeline_of_the_2008_Atlantic_hurricane_season
   and find out how many hurricanes & tropical storms there were for each month from May till
   December 2008. Count a storm that extends into two months by its starting month. For example,
   Gustav is counted for August.

Month	May	Jun	Jul	Aug	Sep	Oct	Nov	Dec
Number of hurricanes & tropical storms								

Add that data to the line graph on the previous page. Also, answer the questions:

- Was the 2008 Atlantic hurricane season more or less active than the season in 2005? In 2007?

- Which month of the 2008 season was the most active month?

- Is the same month also the most active month for 2005, 2006, and 2007 seasons?

   (You can also extend this project to include 2009, 2010, and so on, up to the year prior to doing these exercises.)

4. The table shows Anna's and Alex's test scores in five science tests.

   a. Draw a double-line graph of the data. Note the legend that is already given.

   b. Describe Anna's performance over time. (Did she improve? Get worse? Stay about the same?)

   c. In which test was the difference between Anna's and Alex's point count the greatest?

   In which test was it the smallest?

	Anna	Alex
Test 1	65	72
Test 2	62	66
Test 3	77	71
Test 4	85	59
Test 5	82	68

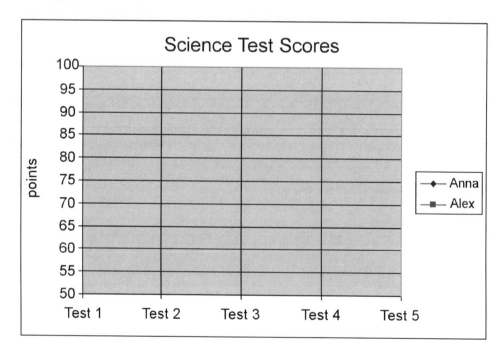

Making Bar Graphs

Bar graphs are used if the data can be separated into distinct groupings or categories. For example, if you study children's eye color, the categories are "blue," "green," "brown," "hazel," *etc.*

The graph on the right shows the number of U.S. households that own a dog, cat, bird, or a horse. A household owning, say, both a dog and a cat would be included in both numbers. Note that the vertical axis scale is in million households.

Notice how the data values are recorded above each bar. To get the real value, multiply that by 1,000,000.

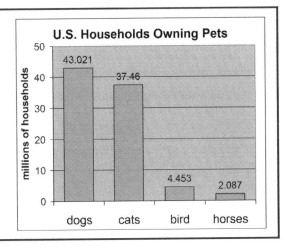

1. According to the graph above, how many U.S. households own a cat? A horse?

2. **a.** Draw the bars into the graph below using the data on the right. Notice that *you* need to figure out the scaling for the horizontal axis (miles).
   Hint: Make sure the largest number in the river lengths fits on the grid, and that there isn't lots of "empty space" left over beyond that.

   b. About how many times longer is the Mississippi-Missouri than the Ohio-Allegheny?

   c. About how many times longer is the Mississippi-Missouri than the Yukon?

River	Length (miles)
Mississippi-Missouri	3,902
Yukon	1,980
Rio Grande	1,900
Columbia	1,450
Colorado	1,450
Ohio - Allegheny	1,306
Snake	1,038

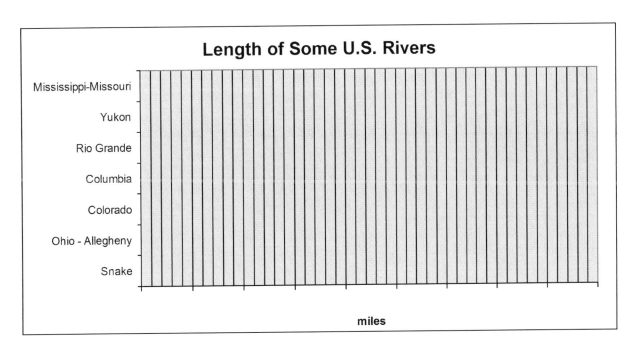

3. The table lists the number of U.S. households that own certain exotic pets.

a. Make a bar graph from the data. Notice that the vertical axis is scaled for *thousands* of households.

b. Estimate the number of households that own either a turtle, a lizard, or a snake.

c. Which is more popular: to own a hamster, a guinea pig, or a gerbil *or* to own a turtle, a lizard, or a snake? Justify your answer.

Pet	Number of Households (in 1,000)
Rabbits	1,870
Turtles	1,106
Hamsters	826
Lizards	719
Guinea Pigs	628
Ferrets	505
Snakes	390
Gerbils	187

Source: 2007 U.S. Pet Ownership & Demographics Sourcebook

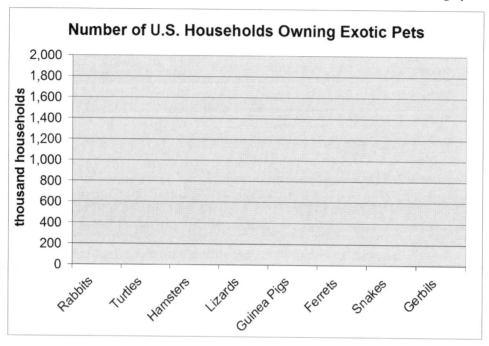

4. Serena asked 20 people in her class how many brothers and sisters they had. Here is her data:
0 3 2 1 1 0 1 1 2 2 1 2 5 3 2 6 1 1 0 2 (Each number is one person's response.)
Draw a bar graph. First count how many people had zero siblings, how many had one sibling, and so on (the frequencies).

Number of siblings	frequency
0	
1	
2	
3	
4	
5	
6	

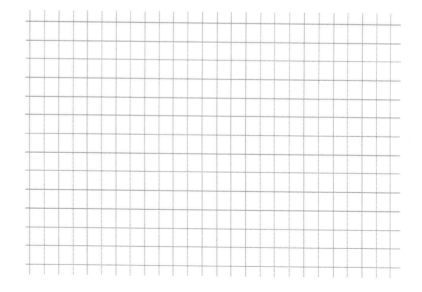

Making Histograms

Histograms are like bar graphs, but the bars are drawn so they touch each other. Histograms are used with numerical data. For an example, let's look at the math test scores of a fifth grade class:

13 40 32 38 32 28 21 30 45 17 22 26 33 25 27 36 42 19 21

First we need to make categories or "bins" for this data, and after that draw a histogram.

1. Let's make *five* categories or bins for the test score data. These bins are shown on the right. The first bin is from 12 to 18 points, the second bin is from 19 to 25 points, and so on.

    How did we come up with those limits for the bins?

    First, we find the <u>smallest</u> and the <u>greatest</u> value among the data. Those are 13 and 45. The first bin has to include 13 and the last bin has to include 45.

    If we want five bins, we find the difference between those numbers and divide that by 5. The result will give us the **width** of each bin.

point count	frequency
12-18	
19-25	
26-32	
33-39	
40-46	

We get $45 - 13 = 32$ and $32 \div 5 = 6.4$. The width could be 6.4 points. However, often it is more reasonable to use a whole number for the bin width, so instead of 6.4, let's make the bins 7 points "wide." We can start the first category at 13 (the lowest score) or even a little bit before that, at 12. The important thing is that the last bin has to be able to include 45, our highest number.

Each bin starts at 7 points higher than the previous one: the second at 19, the third at 26, and so on.

Next, count *how many* individual test scores "fall into," or belong in, each bin—that is the **frequency**. Now you are ready to draw the histogram! Draw it so that the bars touch each other without leaving gaps in between.

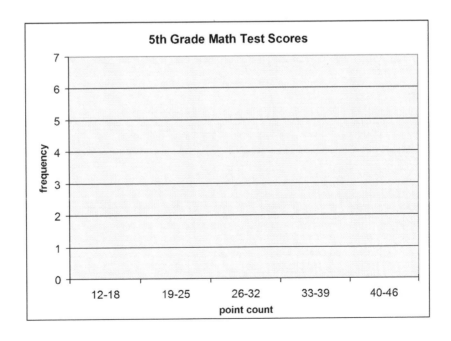

2. The following data describe the weights of 15 healthy female German Shepherd dogs (in pounds).
   65 72 62 60 66 67 65 73 70 64 66 63 68 58 63
   Draw a histogram. Make four categories. First determine the bin width: find the largest and the
   smallest numbers in the data, calculate the difference between them, divide that difference by four,
   and round the result up to the next whole number.

weight	frequency

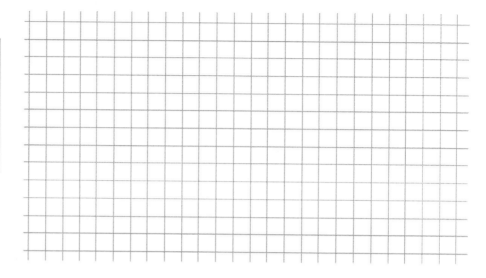

3. Researchers determined the age of 26 African elephants, living in three herds, by their molars.
   Here is the data (each number is the age of one elephant):
   3 0 6 23 12 0 1 15 9 8 43 2 4 10 22 38 5 17 3 8 18 27 19 7 4

   Draw a histogram. Make five categories. Determine the bin width for the categories as above.

age	frequency

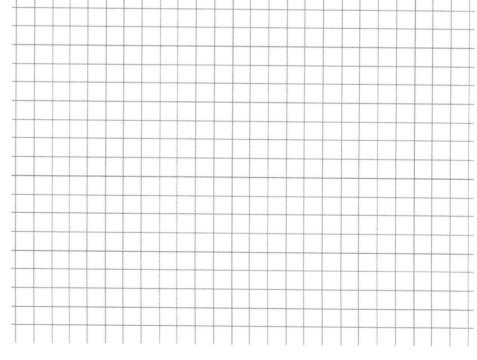

Double Bar Graphs

Double-bar graphs are used to compare two sets of data. Both data sets have to have the same categories.

For example, the chart here shows how many students are able to swim in a certain elementary school. The categories are the five different grades. There is a separate bar for boys and girls for each grade.

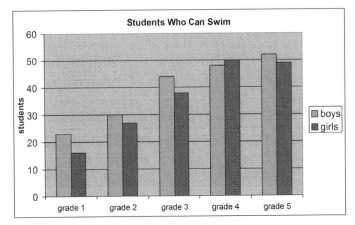

We can see the following:

- In grades 1, 2, and 3, more boys are able to swim than girls.

- In grades 4 and 5 about the same number of girls can swim as boys.

- As students get older, more students are able to swim. This is probably because the school has swimming instruction in grades 2, 3, and 4.

1. **a.** Write in the table how many students can swim and how many cannot swim for grades 1 to 5.

Grade	can swim	cannot swim

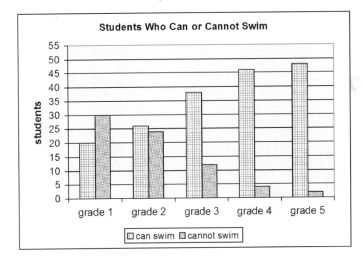

b. How does the number of students who can swim change during the grades 1 - 5?

c. How does the number of students who cannot swim change during the grades 1 - 5?

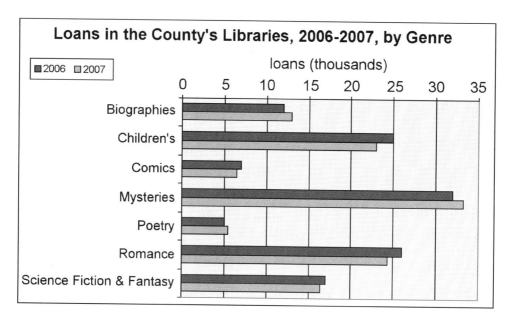

Loans in the County's Libraries, 2006-2007, by Genre

2. **a.** List the genres which had an increase
in the number of loans from 2006 to 2007.

b. Estimate the total number of loans for the three most popular genres in 2006.

c. Estimate the total number of loans for the three least popular genres in 2007.

3. A researcher asked some people over 50 years of age and some people between 30 and 50 years of age about their favorite type of books.

Favorite book genres by age

	over 50	30-50
Biographies & Memoirs	245	36
Comics	45	126
Mysteries	357	186
Poetry	56	22
Romance	128	215
Science Fiction & Fantasy	37	267

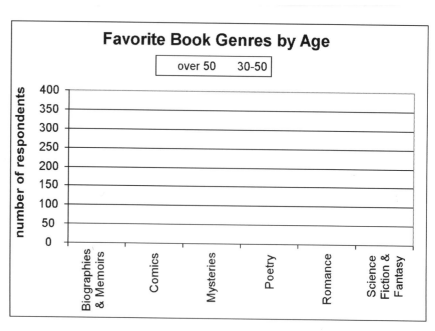

Favorite Book Genres by Age

a. Draw a double-bar graph of the data. You will also need to fill in the legend.

b. Find the genre that holds the last place in the one age
group while holding the first place in the other age group.

Average (Mean)

Example 1. Five children earned these amounts of money for a job: $12, $27, $18, $9, and $22. The graph below shows visually how much each child earned.

Together, they earned $88. If this $88 had been divided equally among the children, each child would have gotten $18. (Of course it was not, because the children got paid according to how much they worked.)

This $18 is the **average** pay. Average $= \dfrac{\$12 + \$27 + \$18 + \$9 + \$22}{5} = \dfrac{\$88}{5} = \$18.$

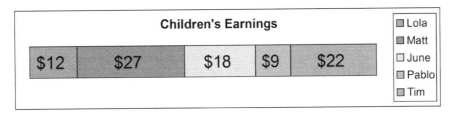

The graph on the bottom shows the situation *if* each child had received the average earning ($18). Notice that $18 is sort of in the "middle" or in between the lowest and highest earnings.

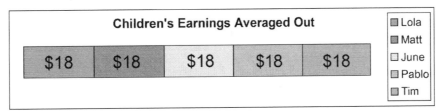

- To calculate the average, first add all the numbers in the data set, and then divide the sum by the number of items in the data set. In other words,

$$\text{average} = \frac{\text{sum of all the items}}{\text{the number of items}}$$

- The average is always somewhere in the middle of a set of data: it is more than the smallest number and less than the largest number of the data.

- The average is also called the **mean**. We will use both terms in this lesson so you get used to both.

1. Calculate the average of the data sets. Do not use a calculator.

   a. 2, 4, 5, 9, 0, 4, 1, 7   **b.** 13, 16, 20, 22, 16, 13, 17, 12, 15

2. Calculate the mean of the data sets to the nearest tenth. This time use a calculator.

   a. 2, 4.3, 5, 9, 4.7, 9.4, 3.7, 5.1   **b.** 312, 288, 284, 329, 293, 302

Average and Line Graphs

Remember that line graphs show how data changes over *time*. Let's look at the muffin sales again.

Muffin Sales, Week 11	
Day	**Muffins sold**
Mon	24
Tue	36
Wed	41
Thu	33
Fri	17

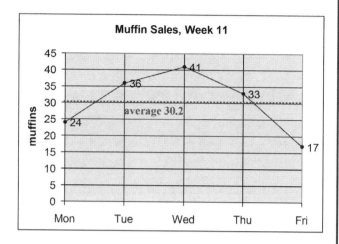

The average is $\dfrac{24 + 36 + 41 + 33 + 17}{5} = 30.2$ muffins per day.

On average, she sold 30.2 muffins or about 30 muffins per day. See how this 30.2 muffins is plotted on the line graph, using a dashed line. Notice that the average is somewhere in the middle of the data: some data points are above, some are below it.

If Mary had sold 30.2 (or 30 1/5) muffins every day, she would have sold the same total amount in five days as she actually did: 151 muffins.

If Mary had sold 30.2 muffins every day, what would the line graph look like?

It would be a horizontal line, with each day's data value at the same level of 30.2.

3. Find the average visitor count to Juanita's blog in the year 2008. Then plot the average in the line graph with a dashed line, like in the example above.

Month	Visitors
Jan	1039
Feb	1230
Mar	1442
Apr	1427
May	1183
Jun	823
Jul	674
Aug	924
Sep	1459
Oct	1540
Nov	1638
Dec	1149

b. Find the average visitor count for the summer months June through August only.

c. Find the average visitor count for September through December.

4. The average rainfall in the first 15 days of April was 5.067 mm. What was the total rainfall in the first 15 days of April (in millimeters)?

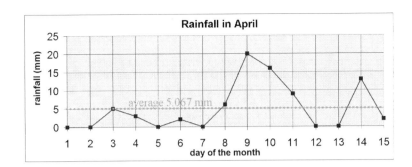

Rainfall in April

5. The birth weights of a certain litter of piglets were:
   1,400 g   1,480 g   1,250 g   1,710 g   1,630 g   1,250 g   1,700 g   1,820 g   1,500 g

   a. Find the average weight to the nearest gram.

   b. How many grams *lighter* than the average were the lightest (two) piglets?

   c. How many grams *heavier* than the average was the heaviest piglet?

   d. Remove the two lightest piglets' weights from the data.
   Now calculate the average again.
   Did the average change? If it did, by how much?

6. The data below gives the monthly salaries of StarMop Inc. employees:
   $1,146 $1,178 $1,189 $1,209 $1,209 $1,210 $1,213 $1,215 $3,400

   a. Calculate the mean.

   b. Remove the person with the highest salary from the data set.
   Calculate the mean again. How did it change?

Joan checked the price of a certain plasma TV in four different stores. In three of the stores the price was $549, $589, and $599. She calculated that the average price was $567.

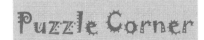

What was the price in the fourth store?
Choose the right answer: **a.** $609   **b.** $531   **c.** $460   **d.** $567

Mean, Mode, and Bar Graphs

Do you think you could calculate the average of the data shown in the bar graph? After all, there are numbers involved.

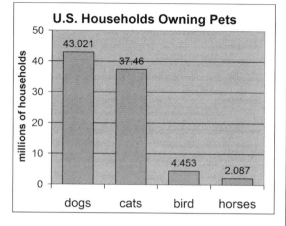

Actually, we cannot. To see why, think *what kind of data* produced this graph originally. What were the people in the study asked? What did they respond?

The people were asked something like, "What pets do you have?" The people would have answered, "cat," "dog," "fish," and similar.

The original data set consists simply of the words "cat," "dog," "bird," and "horse"—each one listed many times, because each word corresponds to the answer of one particular household.

cat, cat, dog, dog, dog, bird, dog, dog, bird, cat, dog, horse, dog, cat, dog....

We cannot calculate anything from this kind of data set because it is **not numerical data**. The only thing we can do is to determine the most commonly occurring item—the **mode**.

In this case, the mode is *dog.* You can see that from the graph: the tallest bar is for dogs.

Mode is the most commonly occurring item in a data set.

- Sometimes a set of data has two or more modes. For example, the data set *green, green, blue, blue, black, brown, hazel* has two modes: both green and blue are equally common.

- If none of the items occurs twice or more, there is no mode. For example, this data *green, blue, pink, red, black, brown, purple* has no mode.

1. Find the mode of the data set shown in the bar graph on the right.

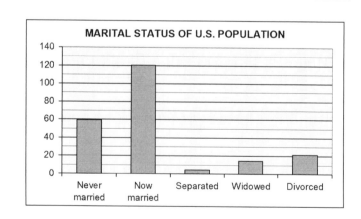

2. **a.** Find the mode of this data:

   water, pop, juice, pop, juice, water, milk, water, pop, pop, juice, pop

   b. If the above words are the answers of 12 people to some question, what could have been the question?

3. Nineteen children were asked about their favorite ice cream flavor. Here are their responses:

strawberry, vanilla, chocolate, vanilla, chocolate chip, chocolate, pecan, pecan, vanilla, vanilla, strawberry, chocolate chip, vanilla, chocolate, chocolate, vanilla, strawberry, chocolate chip, vanilla.

a. Find the mode.

b. Draw a bar graph.

c. If possible, calculate the mean.

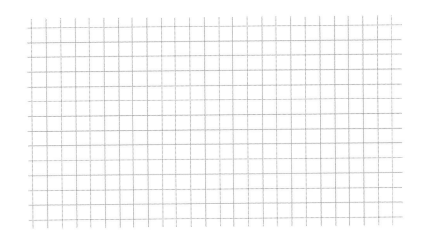

4. These are the spelling test scores of a fifth grade class:
   4 5 7 9 9 10 10 11 11 12 12 12 13 14 17 18 18 18 19 19 19 20 24 25

a. Find the mode.

b. Draw a bar graph.

c. If possible, calculate the mean.

Test Score	Frequency
< 8	
8..10	
11..13	
14..16	
17..19	
20..22	
23..25	

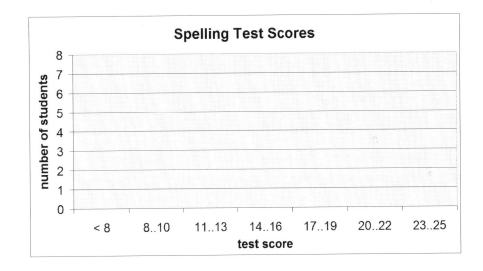

5. **a.** Find the mode.

b. Draw a bar graph.

c. If possible, calculate the average.

d. There were _____ students in all. What *fraction* of the students got grade B?

Grades of a math class

Grade	Frequency
F	3
D	8
C	12
B	17
A	10

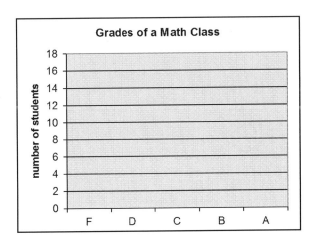

Statistics Project (optional)

Collect information, organize it, draw a graph, and analyze! Choose one or more of the following projects:

1. The temperature of ice water

You need: a glass of room temperature water, an ice cube, and a good thermometer.
Before you start, answer this question: Do you think the temperature will decrease so that your line graph will form a straight line downwards?

First record the temperature of the water in your glass. This should be equal to the room temperature. Drop the ice cube into the glass, and then record the temperature of the water in the ice-water mixture *every two minutes* until the ice cube has completely melted. Better yet, keep recording the temperature until the mixture returns back to room temperature. (This will take quite a bit longer, of course.)

Make a line graph. Are there any surprises in its shape?

If you have access to the Internet, you can also make a line graph online at
http://nces.ed.gov/nceskids/createagraph/default.aspx

2. Your growth

If your parents have records of your growth as a baby or as a child—height, weight, or both—make a line graph of this data. When did you grow the fastest?

If you have access to the Internet, you can also make a line graph online at
http://nces.ed.gov/nceskids/createagraph/default.aspx

3. A survey

Make your own survey. Use questions such as:

What is your favorite color?

a. red   b. blue   c. green   d. yellow
e. purple   f. other, what? _____

After you have answers from 20 of your friends, organize them in a frequency table. Make a bar graph. Find the mode. You will need a separate frequency table and bar graph for each question. See the example on the right.

Question 1	
Choices	**Frequency**
a	
b	
c	
d	
e	
f	

4. The last digit of phone numbers

Pick three pages from a phone directory, and randomly choose 70 numbers from each page. Record the *last* digit of each phone number. Organize your results in a frequency table. Make a bar graph. Find the mode.

Would you expect each number from 0 to 9 to occur equally often as the last digit? Did that happen?

Mixed Review

1. **a.** Write a number that is 5 thousandths, 2 tenths,
and 8 hundredths more than 1.004.

   b. Write a number that is 3 thousandths and
3 tenths less than 3.411.

2. Figure out what was done in each step. It is either addition or subtraction!

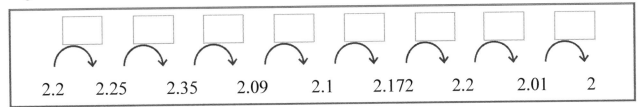

3. Multiply mentally.

a.	b.	c.	d.
$2 \times 0.06 = $ _____	$0.4 \times 0.7 = $ _____	$100 \times 0.12 = $ _____	$1.1 \times 0.9 = $ _____
$2 \times 0.6 = $ _____	$5 \times 0.007 = $ _____	$0.5 \times 0.03 = $ _____	$1000 \times 0.05 = $ _____

4. **a.** Estimate the total
cost in dollars.

   b. Find the total.

   c. Find the error of
estimation.

beans	$4.35
milk	$2.99
dog food	$11.38
broccoli	$2.14
chicken	$7.64

Estimate:

5. Factor the following numbers to their prime factors.

a. 48 / \	b. 71 / \	c. 93 / \

6. Find the value of x.

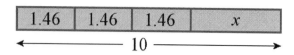

7. Calculate.

a. $2 \times 10^4 =$ _____	**b.** $9 \times 10^6 =$ _____	**c.** $17 \times 10^3 =$ _____

8. Write a division equation where the quotient
   is 210, the divisor is 52, and the dividend is
   unknown. Use a letter for the unknown.
   Then find the value of the unknown.

   _____ ÷ _____ = _____

9. Mark the numbers given in the problem in the bar model.
   Mark what is asked with "?". Then solve the problem.

   Mary and Luisa bought a gift together for $46.
   Mary spent $6 more on it than Luisa.
   How many dollars did each girl spend?

   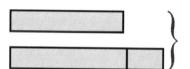

10. With the money John earned from his summer job, he
    paid his phone service for two months ($48 per month),
    spent $120 for a bike, and still had half of his money left.
    How much did he earn?

11. Each story in a tall apartment building is 235 cm high.
    Estimate the total height of the 12-story building, <u>in meters.</u>

12. Divide. Check your answer by multiplying.

a. 38$\overline{)3\ 9\ 5\ 2}$	$\times$ 3 8 $\rule{3cm}{0.4pt}$	**b.** 17$\overline{)2\ 6\ 8\ .\ 6}$	$\times$ $\rule{3cm}{0.4pt}$

13. Divide.

a. Calculate $56 \div 9$ to two decimal digits.	**b.** Change the problem $5.175 \div 0.5$ so that you get a *whole-number divisor*. Then, divide.
$)\rule{4cm}{0.4pt}$	$)\rule{4cm}{0.4pt}$

14. First, estimate the answer by using rounded numbers. Then calculate the exact answer with a calculator. Lastly, find the error of estimation with a calculator.

a. $127{,}285 + 84{,}662$ (round to thousands)	**b.** $12{,}705{,}143 - 6{,}460{,}788$ (round to millions)
My estimation: _____	My estimation: _____
_____	_____
Exact answer: _____	Exact answer: _____
Error of estimation: _____	Error of estimation: _____

Chapter 5 Review

1. Plot the points from the number rule on the coordinate grid. Fill in the rest of the table first, using the rule given.

   The rule is: $y = 9 - x$.

x	0	1	2	3	4
y					

x	5	6	7	8	9
y					

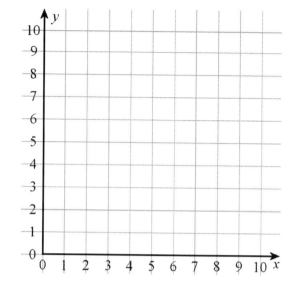

2. Find the mean and mode of this data set to the nearest hundredth: 5, 9, 13, 12, 16, 10, 19, 11, 10.

3. **a.** Estimate what the amount of tractors might have been in the year 2010.

   b. During which decade did the amount of tractors rise the quickest?

   What was the *approximate* amount of increase in tractors during that decade?

   c. Describe the trend in the amount of tractors between 1970 and 1995.

   d. About how many-fold was the increase in tractors between 1930 and 1960?

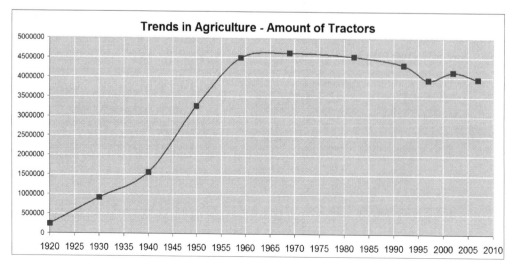

Source: Census of Agriculture

4. A department store was tracking the sales of many items, including umbrellas.

    a. In 2007, in which months were the sales less than 40 umbrellas? How about in 2008?

    b. Find the month with the greatest difference between 2007 and 2008 sales.

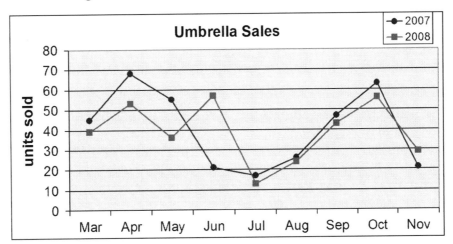

5. Four hundred eight students were asked about how many hours they had slept the previous night.

The results are summarized
in the table below:

Hours of Sleep	Frequency
6	2
7	15
8	56
9	148
10	137
11	40
12	10
total	408

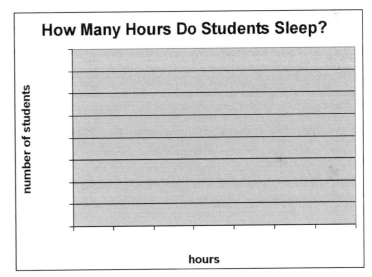

    a. Draw a bar graph. Note that you need to choose the scaling on the vertical axis.

    b. Find the mode.

    c. Which of the following could be a part of the *original* data set?
       (Hint: Think of what was asked and what the students answered.)

       2, 15, 56, 148, 137, 40, 10, 2, 15, …

       6, 10, 8, 8, 9, 7, 11, 10, 9, 10, 11, … .

    d. Which of the following is the average for this data set?

       9.4 hours   8.5 hours   11.1 hours   10.7 hours

Chapter 6: Fractions: Add and Subtract
Introduction

In 5th grade, students study most aspects of fraction arithmetic: addition, subtraction, multiplication, and then in some special cases, division. Division of fractions is studied in more detail in 6th grade. I hope that students have already built a solid conceptual understanding in their minds in previous years, so we can build on that foundation.

The chapter starts out with lessons on various ways to add and subtract mixed numbers. These are meant partially to review and partially to develop speed in fraction calculations. The lesson *Subtracting Mixed Numbers 2* presents an optional way to subtract, where we use a negative fraction. This is only meant for students who can easily grasp subtractions such as $(1/5) - (4/5) = -3/5$, and is not intended to become a "stumbling block." Simply skip the method if your student does not understand it easily.

Students have already added and subtracted *like* fractions in fourth grade. Now it is time to "tackle" the more complex situation of *unlike* fractions.

First, we review how to convert fractions into other equivalent fractions. We begin with a visual model of splitting pieces of pie, and from that, we develop the common procedure for equivalent fractions.

This skill is used immediately in the next lessons about adding and subtracting unlike fractions. We begin this topic by using visual models. From the visual and concrete we gradually advance toward the abstract. Several lessons are devoted to understanding and practicing the basic concept, and also to applying this new skill to mixed numbers.

The lesson *Comparing Fractions* reviews some mental math methods for comparing fractions. Students also learn a "brute force" method based on converting fractions to equivalent fractions. This chapter ends with a lesson on measuring in inches, using units as small as 1/16 of an inch.

The Lessons in Chapter 6

Helpful Resources on the Internet

MIXED NUMBERS

Clara Fraction Ice Cream Shop
Convert improper fractions to mixed numbers, while scooping various ice cream flavors onto the cone.
http://mrnussbaum.com/clarafraction/

Fraction Models
Explore improper fractions, mixed numbers, decimals, and percentages using several models: bar, area, pie, and set. Adjust numerators and denominators to see how they alter the models.
http://illuminations.nctm.org/Activity.aspx?id=3519

Fractions Workshop
Choose "Add mixed fractions with like denominators" and the number of problems you would like to do.
http://mrnussbaum.com/fractions-workshop-ipad.html

Subtracting Mixed Fractions Quiz (Like Denominators)
Drag and drop each answer to the corresponding subtraction problem.
http://www.fractions4kids.com/subtracting-mixed-fractions-quiz/

Subtracting Mixed Numbers with Borrowing
Learn how to borrow mixed fractions with this animation.
https://www.wisc-online.com/learn/formal-science/mathematics/abm701/subtracting-mixed-number-fractions-with-borro

EQUIVALENT FRACTIONS

Equivalent Fractions
You are given a fraction that is shown with a visual model and on a number line, and you need to construct two *other* fractions that are equivalent to the given fraction. Drag two sliders to choose the denominators for your fractions and then click pieces to color them.
http://illuminations.nctm.org/Activity.aspx?id=3510

Fresh Baked Fractions
Practice equivalent fractions by clicking on a fraction that is not equal to others.
http://www.funbrain.com/fract/

Triplets: Equivalent Fractions
Sort the space teams by equivalent fractions to make sure all the athletes get to the correct starting place before the games begin.
https://www.mathplayground.com/Triplets/index.html

Fraction Dolphins
Click on the dolphin with the correct equivalent fraction to the fraction on the bucket of fish.
http://mrnussbaum.com/fraction-dolphins-ipad.html

Fraction Worksheets: Equivalent Fractions with Visual Models
Create custom-made worksheets for equivalent fractions. Choose to include pie images or not.
http://www.homeschoolmath.net/worksheets/equivalent_fractions.php

Fraction Worksheets: Equivalent Fractions, Simplifying, Convert to Mixed Numbers
Create custom-made worksheets for these fraction operations.
http://www.homeschoolmath.net/worksheets/fraction-b.php

ADDITION AND SUBTRACTION

Fraction Videos 1: Addition and Subtraction
A set of videos by the author that cover topics in this chapter.
http://www.mathmammoth.com/videos/fractions_1.php

Adding Fractions with Uncommon Denominators Tool at Conceptua Fractions
A tool that links a visual model to the procedure of adding two unlike fractions.
https://www.conceptuamath.com/app/tool/adding-fractions-with-uncommon-denominators

Add Unlike Fractions with Number Line Models
Practice adding unlike fractions. Click "EXPLAIN" to see a visual illustration and the answer.
http://www.visualfractions.com/AddUnlike/

Drop Zone
Practice making a sum of one using fractions in this interactive online activity.
https://www.brainpop.com/games/dropzone/

Add Mixed Numbers with Unlike Denominators - Quiz
Use this simple online quiz for extra practice.
http://www.mathgames.com/skill/5.72-add-mixed-numbers-with-unlike-denominators

Fruit Shoot Fractions
This game practices addition of fractions. There are several different levels to choose from.
http://www.sheppardsoftware.com/mathgames/fractions/FruitShootFractionsAddition.htm

Fruit Splat
Practice finding the least common denominator. This game has three different levels to choose from.
http://www.sheppardsoftware.com/mathgames/fractions/LeastCommonDenomimator.htm

Fraction Word Problems
Practice adding and subtracting fractions with these interactive word problems.
http://mrnussbaum.com/grade5standards/568-2/

Math Balloons: Fractions
Answer whether the fraction additions are true or false in this timed activity.
http://www.mathnook.com/math/math-balloons-fractions.html

Fraction Bars Blackjack
The computer gives you two fraction cards. You have the option of getting more or "holding". The object is to get as close as possible to 2, without going over, by adding the fractions on your cards.
http://fractionbars.com/Fraction_Bars_Black_Jack/

Old Egyptian Fractions
Puzzles to solve: add fractions like a true Old Egyptian Math Cat!
http://www.mathcats.com/explore/oldegyptianfractions.html

Fraction Worksheets: Addition, Subtraction, Multiplication, and Division
Create custom-made worksheets for the four operations with fractions and mixed numbers.
http://www.homeschoolmath.net/worksheets/fraction.php

ORDERING AND COMPARING

Comparing Fractions Tool at Conceptua Fractions
An interactive tool where students place numbers, visual models, and decimals on a number line.
http://www.conceptuamath.com/app/tool/comparing-fractions

Comparison Shoot Out
Choose level 2 or 3 to compare fractions and shoot the soccer ball to the goal.
http://www.fuelthebrain.com/games/comparison-shootout/

Comparing Fractions—XP Math
Simple timed practice with comparing two fractions.
http://xpmath.com/forums/arcade.php?do=play&gameid=8

Visual Fractions Game
Find a fraction between two given fractions with the help of this visual tool.
http://www.mathplayground.com/visual_fractions.html

Fractional Hi Lo
The computer has selected a fraction. You make guesses and it tells if your guess was too high or low.
http://www.theproblemsite.com/games/hilo.asp

My Closest Neighbor
A neat card game where you need to make a fraction that is as close as possible to the given fraction.
https://denisegaskins.com/2014/08/06/fraction-game-my-closest-neighbor/

Comparing/Ordering Fractions Worksheets
Create customizable worksheets for comparing or ordering fractions. You can include pie images.
http://www.homeschoolmath.net/worksheets/comparing_fractions.php

MEASURING & GENERAL

Measure It!—Practice measuring lines in inches.
https://www.funbrain.com/measure/

Measuring—Practice measuring with a virtual ruler. Choose the category "Inches, Sixteenths".
http://www.abcya.com/measuring.htm

Sal's Sub Shop—Cut the subs to the given measurements.
http://mrnussbaum.com/sal/

Fraction Word Problems
This is a set of 10 interactive word problems with multiple-choice answers involving mixed numbers.
http://mrnussbaum.com/grade5standards/572-2

Who Wants Pizza?—A tutorial and interactive exercises about fraction addition and multiplication.
http://math.rice.edu/~lanius/fractions/

Fraction Lessons—Tutorials, examples, and videos explaining all the basic fraction topics.
http://www.mathexpression.com/learning-fractions.html

Online Fraction Calculator
http://www.homeschoolmath.net/worksheets/fraction_calculator.php

Fraction Terminology

As we study fractions and their operations, it is important that you understand the terms, or words, that we use. This page is for reference. You can even post it on your wall or make your own fraction poster based on it.

$\frac{3}{11}$ The top number is the **numerator**. It *enumerates,* or numbers (counts), *how many* pieces there are.
The bottom number is the **denominator**. It *denominates,* or names, *what kind* of parts they are.

A mixed number has two parts: a whole-number part and a fractional part.

For example, $2\frac{3}{7}$ is a mixed number. Its whole-number part is 2, and its fractional part is $\frac{3}{7}$.

The mixed number $2\frac{3}{7}$ actually means $2 + \frac{3}{7}$.

Like fractions have the same denominator. They have the same kind of parts.

It is easy to add and subtract like fractions, because all you have to do is look at *how many* of that kind of part there are.

$\frac{2}{9}$ and $\frac{7}{9}$ are like fractions.

Unlike fractions have a different denominator. They have different kinds of parts.

It is a little more complicated to add and subtract unlike fractions. You need to first change them into like fractions. Then you can add or subtract them.

$\frac{2}{9}$ and $\frac{3}{4}$ are unlike fractions.

A proper fraction is a fraction that is less than 1 (less than a whole pie). 2/9 is a proper fraction.

$\frac{2}{9}$ is a proper fraction.

An improper fraction is more than 1 (more than a whole pie). It is a *fraction*, so it is written as a fraction and *not* as a mixed number.

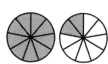

$\frac{11}{9}$ is an improper fraction.

Equivalent fractions are equal in value. If you think in terms of pies, they have the same amount of "pie to eat," but they are written using different denominators, or are "cut into different kinds of slices."

$\frac{3}{9}$ and $\frac{1}{3}$ are equivalent fractions.

Simplifying or reducing a fraction means that, for a given fraction, you find an equivalent fraction that has a "simpler," or smaller, numerator and denominator. (It has fewer but bigger slices.)

$\frac{9}{12}$ simplifies to $\frac{3}{4}$.

Review: Mixed Numbers

This lesson should be mostly review. However, please don't go on to the lessons about adding and subtracting mixed numbers until you *thoroughly* understand the concepts in this lesson.

1. Write the mixed numbers that these pictures illustrate.

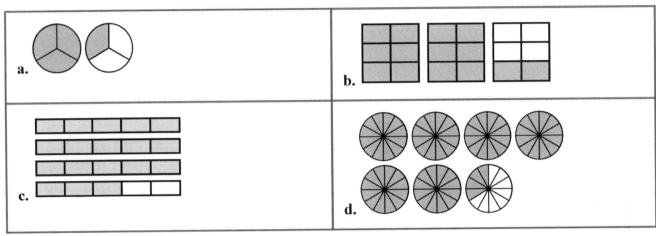

2. Draw pictures of "pies" that illustrate these mixed numbers.

a. $4\frac{2}{3}$

b. $2\frac{3}{5}$

c. $3\frac{2}{6}$

d. $4\frac{7}{8}$

e. $6\frac{8}{10}$

3. Write the mixed number that is illustrated by each number line.

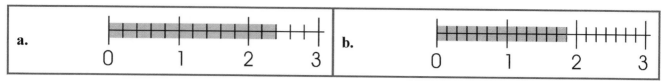

a.

b.

4. Write the fractions and mixed numbers that the arrows indicate.

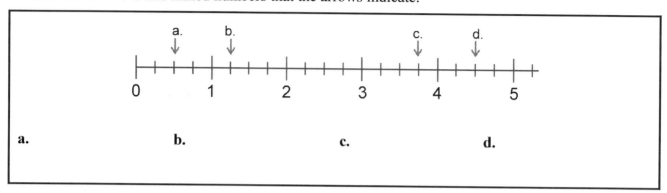

a.   b.   c.   d.

5. Mark the fractions on the number lines.

a. $\dfrac{10}{6}$, $\dfrac{17}{6}$, $\dfrac{12}{6}$, $\dfrac{5}{6}$, $\dfrac{14}{6}$

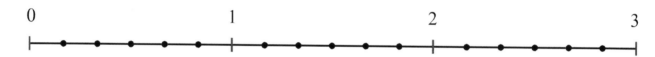

b. $\dfrac{9}{8}$, $\dfrac{22}{8}$, $\dfrac{13}{8}$, $\dfrac{24}{8}$, $\dfrac{11}{8}$

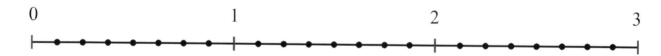

6. **a.** Mark $1\dfrac{3}{5}$ on the number line.   **b.** Write the mixed number that is $\dfrac{4}{5}$ to its right: _____

c. Mark $2\dfrac{1}{5}$ on the number line.   **d.** Write the mixed number that is $\dfrac{3}{5}$ to its left: _____

52

Changing mixed numbers to fractions

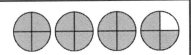

To write $3\frac{3}{4}$ as a fraction, *count* how many fourths there are:

- Each pie has four fourths, so the three complete pies have $3 \times 4 = 12$ fourths.
- Additionally, the incomplete pie has three fourths.
- The total is 15 fourths or 15/4.

Shortcut: $3\frac{3}{4}$ ⤵+   Numerator: $3 \times 4 + 3 = 15$ $= \dfrac{15}{4}$

Denominator: 4

Multiply the whole number times the denominator, then add the numerator. The result gives you the number of fourths, or the numerator, for the fraction. The denominator will remain the same.

7. Write as mixed numbers and as fractions.

a. $1\frac{2}{5} = \dfrac{}{5}$

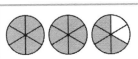

b. $\dfrac{}{} = \dfrac{}{}$

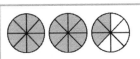

c. $\dfrac{}{} = \dfrac{}{}$

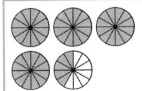

d. $\dfrac{}{} = \dfrac{}{}$

e. $\dfrac{}{} = \dfrac{}{}$

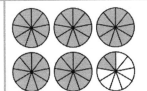

f. $\dfrac{}{} = \dfrac{}{}$

8. May changed $5\frac{9}{13}$ into a fraction, and explained how the shortcut works. Fill in.

There are ____ whole pies, and each pie has ____ slices. So ____ × ____

tells us the number of slices in the whole pies. Then the fractional part 9/13 means that

we add ____ slices to that. In total we get ____ slices, each one a 13th part. So the fraction is

9. Write as fractions. Think of the shortcut.

a. $7\frac{1}{2}$ b. $6\frac{2}{3}$ c. $8\frac{3}{9}$ d. $6\frac{6}{10}$

e. $2\frac{5}{11}$ f. $8\frac{1}{12}$ g. $2\frac{5}{16}$ h. $4\frac{7}{8}$

53

Changing fractions to mixed numbers

To write a fraction, such as $\frac{58}{7}$, as a mixed number, you need to figure out:

- How many *whole* "pies" there are, and
- How many *slices* are left over.

In the case of $\frac{58}{7}$, each whole "pie" will have 7 sevenths. (How do you know?) So we ask:

- How many 7s are there in 58? (Those make the whole pies!)
- After the 7s are gone, how many are left over?

That is solved by the division $58 \div 7$! That division tells you how many 7s there are in 58.

Now, $58 \div 7 = 8$ R2. So you get 8 whole pies, with 2 slices or 2 sevenths left over.

To write that as a fraction, we get $\frac{58}{7} = 8\frac{2}{7}$.

Example 1. $\frac{45}{4}$ is the same as $45 \div 4$, and $45 \div 4 = 11$ R1. So, we get 11 whole pies and 1 fourth-part or slice left over. Writing that as a mixed number, $\frac{45}{4} = 11\frac{1}{4}$.

The Shortcut: Think of the fraction bar as a *division* symbol, and DIVIDE. The quotient tells you the whole number part, and the remainder tells you the numerator of the fractional part.

10. Rewrite the "division problems with remainders" as problems of "changing fractions to mixed numbers."

a. $47 \div 4 = 11$ R3 $$\frac{47}{4} = 11\frac{3}{4}$$	**b.** $35 \div 8 = 4$ R3 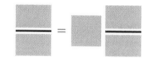	**c.** $19 \div 2 = $ ___ R ___ 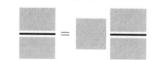
d. $35 \div 6 = $ ___ R ___ 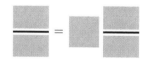	**e.** $72 \div 10 = $ ___ R ___ 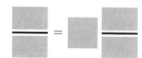	**f.** $22 \div 7 = $ ___ R ___ 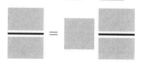

11. Write these fractions as mixed numbers (or as whole numbers, if you can).

a. $\frac{62}{8}$	**b.** $\frac{16}{3}$	**c.** $\frac{27}{5}$	**d.** $\frac{32}{9}$
e. $\frac{7}{2}$	**f.** $\frac{25}{4}$	**g.** $\frac{50}{6}$	**h.** $\frac{32}{5}$
i. $\frac{24}{11}$	**j.** $\frac{39}{3}$	**k.** $\frac{57}{8}$	**l.** $\frac{87}{9}$

Adding Mixed Numbers

You can simply **add the whole numbers and fractional parts separately:**

$$1\frac{1}{7} + 5\frac{3}{7} = 6\frac{4}{7} \qquad \text{or in columns} \rightarrow$$

$$\begin{array}{r} 1\frac{1}{7} \\ + 5\frac{3}{7} \\ \hline 6\frac{4}{7} \end{array}$$

However, often the sum of the fractional parts is more than one whole pie. Study the example below. In it, the sum of the fractional parts is 7/6. Think of that as 1 1/6, and add the 1 to the sum of the whole numbers to get 3. The final answer is 3 1/6.

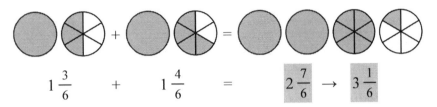

$$1\frac{3}{6} \quad + \quad 1\frac{4}{6} \quad = \quad 2\frac{7}{6} \rightarrow 3\frac{1}{6}$$

1. These mixed numbers have a fractional part that is more than one "pie." Change them so that the fractional part is less than one. The first one is done for you.

 a. $3\frac{3}{2} \rightarrow 4\frac{1}{2}$ **b.** $1\frac{11}{9}$ **c.** $6\frac{7}{4}$ **d.** $3\frac{13}{8}$

2. Write the addition sentences that the pictures illustrate and then add.

a.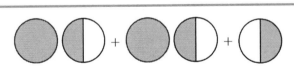

b.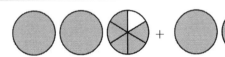

c.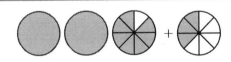

d.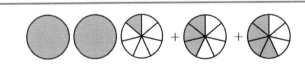

e.

3. Add.

a. $3\frac{2}{3} + 8\frac{1}{3} =$

b. $4\frac{4}{5} + 1\frac{3}{5} =$

c. $6\frac{8}{9} + 1\frac{2}{9} =$

d. $3\frac{6}{7} + 2\frac{4}{7} =$

4. Add.

a. $4\frac{3}{7}$ $+\ 5\frac{5}{7}$ ___ $9\frac{8}{7} \rightarrow 10\frac{1}{7}$	**b.** $3\frac{3}{5}$ $+\ 3\frac{4}{5}$ ___ $\rightarrow$	**c.** $4\frac{6}{9}$ $+\ 2\frac{7}{9}$ ___	**d.** $7\frac{6}{8}$ $+\ 2\frac{7}{8}$ ___

5. Tom has one string that is 7 3/8 inches long and another that is 5 7/8 inches long. He tied them together. In making the knot, he lost 1 4/8 inches from the total length. How long is the combined string now?

6. Gisele found two recipes for corn bread. If she uses recipe 1, she needs to double it.

Which would use more flour, recipe 1 doubled, or recipe 2?

Recipe 1:

1/2 cup cornmeal
3/8 cup wheat flour

(plus other ingredients)

Recipe 2:

1 cup cornmeal
1 cup wheat flour

(plus other ingredients)

How much more?

7. Find the missing addend. Imagine drawing more in the picture.

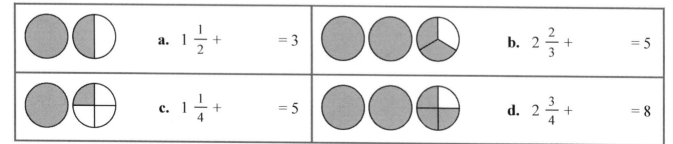

a. $1\frac{1}{2} +$ ____ $= 3$

b. $2\frac{2}{3} +$ ____ $= 5$

c. $1\frac{1}{4} +$ ____ $= 5$

d. $2\frac{3}{4} +$ ____ $= 8$

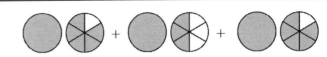

Sometimes the sum of the fractional parts can be two or more whole pies. Just figure out how many whole pies you can make from the fractional parts and add them to the whole-number part.

$$1\frac{5}{6} + 1\frac{3}{6} + 1\frac{5}{6} = 3\frac{13}{6} \rightarrow 5\frac{1}{6}$$

8. Convert these mixed numbers so that the fractional part is less than one.

a. $3\frac{13}{5}$

b. $1\frac{11}{4}$

c. $6\frac{13}{4}$

d. $3\frac{19}{8}$

9. Add these mixed numbers.

a. $3\frac{1}{6} + 2\frac{5}{6} =$

b. $4\frac{4}{5} + 1\frac{2}{5} + 5\frac{2}{5} =$

c. $6\frac{4}{8} + 1\frac{6}{8} + 1\frac{7}{8} =$

d. $3\frac{6}{10} + 3\frac{8}{10} + \frac{9}{10} =$

10. Add the mixed numbers.

| **a.** $10\frac{7}{9}$ $2\frac{5}{9}$ $+\ \ 3\frac{8}{9}$ _____ $\rightarrow$ | **b.** $1\frac{5}{11}$ $3\frac{9}{11}$ $+\ 2\frac{8}{11}$ _____ $\rightarrow$ | **c.** $2\frac{5}{6}$ $5\frac{4}{6}$ $+\ 2\frac{3}{6}$ _____ $\rightarrow$ | **d.** $1\frac{7}{10}$ $\frac{9}{10}$ $+\ 10\frac{6}{10}$ _____ $\rightarrow$ |

11. Jeremy runs 2 ¼ miles four days a week.
Robert runs 3 ½ miles three times a week.
Which boy runs more in one week?
How much more?

12. Find the missing addends.

| **a.** $2\frac{1}{4} + 1\frac{1}{4} + = 5$ | **b.** $3\frac{2}{5} + 2\frac{2}{5} + = 8$ | **c.** $2\frac{1}{3} + \frac{2}{3} + = 4\frac{1}{3}$ |

Subtracting Mixed Numbers 1

Strategy 1: Renaming / regrouping

In this method we cut one of the whole pies into slices, and join these slices with the existing slices. After that, we can subtract. It is the same process as regrouping in subtraction of whole numbers.

Example 1. Solve 3 2/6 − 1 5/6.

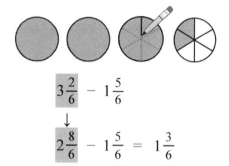

$$3\frac{2}{6} - 1\frac{5}{6}$$

$$\downarrow$$

$$2\frac{8}{6} - 1\frac{5}{6} = 1\frac{3}{6}$$

At first we have three uncut pies and 2/6 more. We cut one of the whole pies into sixths. We end up with only two whole (uncut) pies and 8 sixths.

We say that 3 2/6 has been **renamed** as 2 8/6. Now we can subtract 1 5/6 easily.

We can also solve this problem by writing the mixed numbers one under the other.

$$
\begin{array}{r}
2\ \frac{8}{6} \\
\cancel{3}\ \cancel{\frac{2}{6}} \\
-\ 1\ \frac{5}{6} \\
\hline
1\ \frac{3}{6}
\end{array}
$$

We **regroup** (borrow) 1 whole pie as 6 sixths. There are already 2 sixths in the fractional parts column, so we add the 6/6 and 2/6 and write 8/6 in place of 2/6. Now we can subtract the 5/6.

Example 2. Solve 2 1/8 − 5/8.

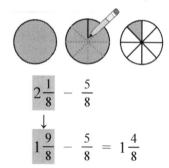

$$2\frac{1}{8} - \frac{5}{8}$$

$$\downarrow$$

$$1\frac{9}{8} - \frac{5}{8} = 1\frac{4}{8}$$

Or:

We need to regroup 1 whole as 8/8. The 8/8 and the existing 1/8 make a total of 9/8.

$$
\begin{array}{r}
1\ \frac{9}{8} \\
\cancel{2}\ \cancel{\frac{1}{8}} \\
-\ 1\ \frac{5}{8} \\
\hline
\frac{4}{8}
\end{array}
$$

1. Do not subtract anything. Divide one whole pie into fractional parts and rename the mixed number.

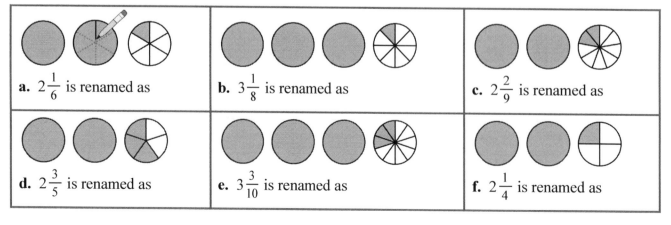

a. $2\frac{1}{6}$ is renamed as

b. $3\frac{1}{8}$ is renamed as

c. $2\frac{2}{9}$ is renamed as

d. $2\frac{3}{5}$ is renamed as

e. $3\frac{3}{10}$ is renamed as

f. $2\frac{1}{4}$ is renamed as

2. Rename, then subtract. Be careful. Use the pie pictures to check your calculation.

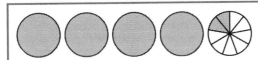

a. $4\frac{2}{9} - 1\frac{8}{9}$

↓

$= 3\frac{11}{9} - 1\frac{8}{9} =$

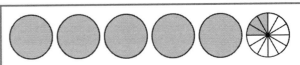

b. $5\frac{3}{12} - 2\frac{7}{12}$

↓

$= 4\frac{}{12} - 2\frac{7}{12} =$

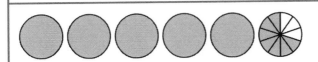

c. $5\frac{7}{10} - 3\frac{9}{10}$

↓

$= \frac{}{} - 3\frac{9}{10} =$

d. $4\frac{3}{8} - 1\frac{7}{8}$

↓

$= \frac{}{} - 1\frac{7}{8} =$

3. Regroup (if necessary) and subtract.

a.
$$\begin{array}{r} 2\frac{}{9} \\ \cancel{3}\ \cancel{\frac{4}{9}} \\ -\ \ \frac{8}{9} \\ \hline \end{array}$$

b.
$$\begin{array}{r} 7\ \frac{4}{9} \\ -\ 2\ \frac{7}{9} \\ \hline \end{array}$$

c.
$$\begin{array}{r} 12\ \frac{9}{12} \\ -\ 6\ \frac{11}{12} \\ \hline \end{array}$$

d.
$$\begin{array}{r} 8\ \frac{3}{14} \\ -\ 5\ \frac{9}{14} \\ \hline \end{array}$$

e.
$$\begin{array}{r} 14\ \frac{7}{9} \\ -\ 3\ \frac{5}{9} \\ \hline \end{array}$$

f.
$$\begin{array}{r} 11\ \frac{5}{21} \\ -\ 7\ \frac{15}{21} \\ \hline \end{array}$$

g.
$$\begin{array}{r} 26\ \frac{4}{19} \\ -\ 14\ \frac{15}{19} \\ \hline \end{array}$$

h.
$$\begin{array}{r} 10\ \frac{3}{20} \\ -\ 5\ \frac{7}{20} \\ \hline \end{array}$$

Strategy 2: Subtract in Parts

First, subtract what you can from the fractional part of the minuend. Then subtract the rest from one of the whole pies. Study the examples.

Example 3. $2\frac{1}{8} - \frac{5}{8}$

$= 2\frac{1}{8} - \frac{1}{8} - \frac{4}{8}$

$= \quad 2 \quad - \frac{4}{8} = 1\frac{4}{8}$

First we take away only 1/8, which leaves 2 whole pies. Then we subtract the rest (4/8) from one of the whole pies

Example 4. $3\frac{2}{9} - 2\frac{7}{9}$

$= 3\frac{2}{9} - 2\frac{2}{9} - \frac{5}{9}$

$= \quad 1 \quad - \frac{5}{9} = \frac{4}{9}$

We cannot subtract 7/9 from 2/9. So, first we subtract 2 and 2/9, which leaves 1 whole pie. The rest, 5/9, is subtracted from the last whole pie.

4. Subtract in parts. Remember: you can *add* to check a subtraction problem.

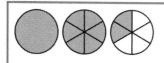

a. $2\frac{2}{6} - \frac{5}{6} =$

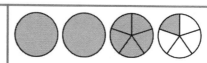

b. $3\frac{1}{5} - 2\frac{3}{5} =$

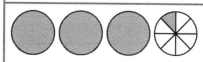

c. $3\frac{1}{8} - 1\frac{7}{8} =$

d. $3\frac{2}{7} - 2\frac{6}{7} =$

e. $5\frac{2}{9} - \frac{5}{9} - 1\frac{8}{9} =$

5. Subtract in two parts. Write a subtraction sentence.

a. Cross out $\frac{3}{4}$.

b.  Cross out $1\frac{5}{7}$.

c. 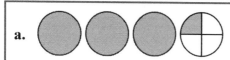 Cross out $1\frac{5}{9}$.

d.  Cross out $1\frac{11}{12}$.

Example 5. Look at Mia's math work: $7\frac{1}{6} - 2\frac{5}{6} = 9\frac{6}{6} = 10$. Can you see why it is wrong?

If you have 7 and a bit and you subtract 2 and some, you cannot get 10 as an answer! In reality, Mia was *adding* instead of subtracting. (If you have ever done that, you are not alone—it is a common error.)

Always check if your answer is reasonable.

6. Subtract.

a. $8\frac{1}{5} - 3\frac{3}{5} =$	**b.** $4\frac{2}{8} - 1\frac{7}{8} =$	**c.** $12\frac{4}{13} - 9\frac{8}{13} =$
d. $11\frac{2}{15} - 6\frac{6}{15} =$	**e.** $7\frac{1}{20} - 3\frac{7}{20} =$	**f.** $6\frac{14}{100} - 2\frac{29}{100} =$

7. Two sides of a triangle measure 3 5/8 in, and the perimeter of the triangle is 10 1/8 in. How long is the third side of the triangle?

8. Ellie had 4 yards of material. She needed 7/8 yard for making a skirt, and she made two. How much material is left?

9. Harry wants to bake chocolate chip cookies. The recipe calls for 1 3/4 cups of flour and he is making a double batch. However, Harry *only* has 3/4 cup of flour! How much flour would Harry need to borrow from his neighbor to have enough to make the cookies?

Puzzle Corner Subtract. The pies may help.

$2\frac{1}{2} - 1\frac{2}{3} =$

Subtracting Mixed Numbers 2

Strategy 3: Use a negative fraction (optional; you can skip this)

Subtract the whole numbers and the fractional parts separately. You may get a *negative* fraction. Treat that as another subtraction problem, and subtract once more to get the final answer.

Example 1. $6\frac{2}{10} - 2\frac{5}{10} = ?$	**Example 2.** $8\frac{1}{7} - 5\frac{6}{7} = ?$
Subtract $6 - 2 = 4$ (the whole numbers), and $\frac{2}{10} - \frac{5}{10} = -\frac{3}{10}$ (the fractions). Notice we get a negative fraction. Lastly, we combine the two results, 4 and −3/10, to get $4 - \frac{3}{10} = 3\frac{7}{10}$.	Subtract $8 - 5 = 3$ (the whole numbers), and $\frac{1}{7} - \frac{6}{7} = -\frac{5}{7}$ (the fractions). Notice we get a negative fraction. Lastly, we combine the two results, 3 and −5/7, to get $3 - \frac{5}{7} = 2\frac{2}{7}$.

1. Subtract using any strategy.

a. $5\frac{3}{8} - 1\frac{7}{8} =$	**b.** $9\frac{2}{15} - 5\frac{8}{15} =$
c. $7\frac{11}{30} - 4\frac{9}{30} =$	**d.** $4\frac{5}{8} - 2\frac{7}{8} =$
e. $13\frac{2}{5} - 4\frac{2}{5} =$	**f.** $16\frac{5}{12} - 4\frac{11}{12} =$

2. You have 3 ¾ kg of ground beef. Your neighbor buys ¾ kg of it and you use ¾ kg to make meatballs. How much beef do you have left?

3. Subtract. Then write an addition that matches with the subtraction.

a. $5\frac{1}{11} - 3\frac{2}{11} =$	**b.** $6\frac{6}{7} - 1\frac{5}{7} =$	**c.** $6\frac{2}{15} - 1\frac{9}{15} =$

4. Find the missing minuend or subtrahend.

a. $ - 2\frac{1}{5} = 3\frac{2}{5}$	b. $ - 2\frac{6}{12} = 3\frac{5}{12}$	c. $7\frac{8}{9} - = 4\frac{1}{9}$
d. $5 - = 2\frac{2}{3}$	e. $ - 5\frac{7}{12} = 1\frac{11}{12}$	f. $5\frac{1}{6} - = 2\frac{5}{6}$

5. Subtract using any strategy. Color the answer squares as indicated.

a. (yellow) $\qquad 5\frac{2}{9} - 2\frac{7}{9}$

b. (blue) $\qquad 7\frac{8}{15} - 4\frac{11}{15}$

c. (blue) $\qquad 5\frac{6}{11} - 3\frac{2}{11}$

d. (yellow) $\qquad 4\frac{1}{9} - 2\frac{3}{9}$

e. (green) $\qquad 17\frac{2}{9} - 4\frac{5}{9}$

f. (green) $\qquad 5\frac{1}{11} - 3\frac{9}{11}$

g. (green) $\qquad 10\frac{1}{12} - 4\frac{7}{12}$

h. (yellow) $\qquad 4\frac{3}{10} - 2\frac{3}{10}$

i. (green) $\qquad 5\frac{1}{10} - 3\frac{9}{10}$

j. (yellow) $\qquad 8\frac{1}{8} - 2\frac{5}{8}$

k. (blue) $\qquad 7\frac{1}{11} - 3\frac{5}{11}$

l. (blue) $\qquad 9\frac{7}{8} - 3\frac{1}{8}$

m. (yellow) $\qquad 15\frac{3}{12} - 10\frac{4}{12}$

$4\frac{11}{12}$	$1\frac{2}{10}$	$12\frac{6}{9}$	$2\frac{4}{9}$
$3\frac{7}{11}$			$2\frac{4}{11}$
$2\frac{12}{15}$		2	$6\frac{6}{8}$
$1\frac{7}{9}$	$1\frac{3}{11}$	$5\frac{6}{12}$	$5\frac{4}{8}$

6. Add and subtract.

a. $2\frac{1}{4} + 5\frac{3}{4} - 3\frac{2}{4} =$	b. $4\frac{5}{6} + 6\frac{3}{6} - 1\frac{4}{6} =$
c. $9\frac{3}{8} + 2\frac{7}{8} - 3\frac{6}{8} =$	d. $7\frac{7}{12} + 3\frac{11}{12} - 1\frac{2}{12} =$

Equivalent Fractions 1

These two fractions are **equivalent fractions** because they picture the same amount. You could say that you get to "eat" the same amount of "pie" either way.

$$\frac{3}{4} = \frac{6}{8}$$

In the second picture, **each slice** has been **split or cut into two pieces**. The arrows show into how many new pieces each piece was split.

$$\frac{1}{3} = \frac{4}{12}$$

Each slice has been split into four.

BEFORE: 1 colored piece, 3 total.
AFTER: 4 colored pieces, 12 total.

Notice that we get *four* times as many colored pieces and *four* times as many total pieces. This means that both the numerator and the denominator get multiplied by 4.

When all of the pieces are split the same way, both the number of colored pieces (the numerator) and the total number of pieces (the denominator) get multiplied by the same number.

1. Connect the pictures that show equivalent fractions. Write the name of each fraction beside its picture.

 $\frac{1}{2}$

2. Make a chain of equivalent fractions.

 = = =

$$\frac{1}{2} = \quad = \quad = \quad = \frac{5}{} = \frac{6}{} = \quad = $$

3. Split the pieces by drawing the new pieces in the right-hand picture. Write the equivalent fractions.

a. Split each piece <u>in two</u>.	b. Split each piece <u>into three</u>.	c. Split each piece <u>in two</u>.
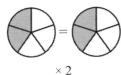 $\times 2$ $\frac{2}{5} = \frac{}{}$ $\times 2$	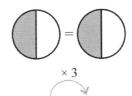 $\times 3$ $\frac{1}{2} = \frac{}{}$ $\times 3$	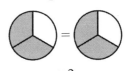 $\times 2$ $\frac{2}{3} = \frac{}{}$ $\times 2$
d. Split each piece <u>in two</u>.	e. Split each piece <u>into three</u>.	f. Split each piece <u>in two</u>.
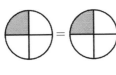 $=$	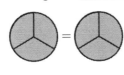 $=$	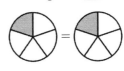 $=$
g. Split each piece <u>in two</u>.	h. Split each piece <u>in two</u>.	i. Split each piece <u>into five</u>.
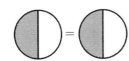 $=$	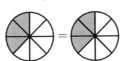 $=$	 $=$

4. Write the equivalent fraction. Imagine or draw the helping arrows.

a. Split each piece into four.	b. Split each piece in two.	c. Split each piece into six.	d. Split each piece into four.	e. Split each piece into five.
$\frac{3}{4} = \frac{\square}{\square}$	$\frac{5}{8} = \frac{\square}{\square}$	$\frac{1}{2} = \frac{\square}{\square}$	$\frac{2}{7} = \frac{\square}{\square}$	$\frac{1}{4} = \frac{\square}{\square}$
f. Split each piece into three.	g. Split each piece into ten.	h. Split each piece into eight.	i. Split each piece into seven.	j. Split each piece into eight.
$\frac{2}{7} = \frac{\square}{\square}$	$\frac{5}{8} = \frac{\square}{\square}$	$\frac{1}{2} = \frac{\square}{\square}$	$\frac{3}{5} = \frac{\square}{\square}$	$\frac{3}{7} = \frac{\square}{\square}$

5. Figure out how many ways the pieces were split and write the missing numerator or denominator.

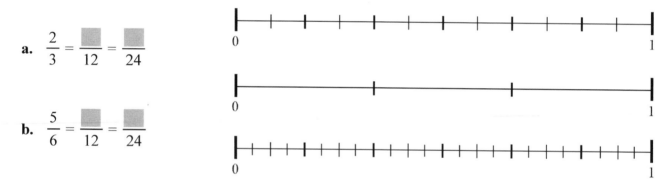

a. Pieces were split into <u>three</u>.	**b.** Pieces were split into ____ .	**c.** Pieces were split into ____ .	**d.** Pieces were split into ____ .	**e.** Pieces were split into ____ .
$\dfrac{4}{7} = \dfrac{}{21}$ ×3 / ×3	$\dfrac{4}{5} = \dfrac{}{20}$	$\dfrac{1}{6} = \dfrac{}{18}$	$\dfrac{6}{7} = \dfrac{}{14}$	$\dfrac{2}{3} = \dfrac{8}{}$
f. $\dfrac{7}{10} = \dfrac{14}{}$	**g.** $\dfrac{5}{9} = \dfrac{15}{}$	**h.** $\dfrac{1}{8} = \dfrac{6}{}$	**i.** $\dfrac{4}{9} = \dfrac{}{54}$	**j.** $\dfrac{8}{11} = \dfrac{}{44}$
k. $\dfrac{3}{10} = \dfrac{9}{}$	**l.** $\dfrac{2}{11} = \dfrac{6}{}$	**m.** $\dfrac{4}{7} = \dfrac{}{56}$	**n.** $\dfrac{1}{6} = \dfrac{}{54}$	**o.** $\dfrac{7}{8} = \dfrac{}{64}$

6. Mark the equivalent fractions on the number lines.

a. $\dfrac{2}{3} = \dfrac{}{12} = \dfrac{}{24}$

0 ⊢——————————————————————┤ 1

b. $\dfrac{5}{6} = \dfrac{}{12} = \dfrac{}{24}$

0 ⊢——————————————————————┤ 1

0 ⊢——————————————————————┤ 1

c. Find and mark two fractions on the 12th parts number line that do *not* have an equivalent fraction on the 3rd parts number line. Write them here →

d. Find and mark two fractions on the 24th parts number line that do *not* have an equivalent fraction on the 12th parts number line. Write them here →

7. A family of four baked two pizzas. Dad ate 1/2 of one pizza, Mom and Cindy ate 1/3 of a pizza each, and Derek ate the rest.

a. Write equivalent fractions for the fractions mentioned in the problem, using 1/12 parts.

b. Figure out which fraction of a pizza Derek ate.

	Fraction	Equivalent fraction
Dad	1/2	
Mom		
Cindy		
Derek		

66

Equivalent Fractions 2

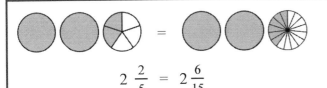

$$2\frac{2}{5} = 2\frac{6}{15}$$

Here you see the mixed number 2 2/5 changed into an equivalent mixed number 2 6/15. Actually, we only changed the fractional part, 2/5, into the equivalent fraction 6/15 and obviously the whole-number part did not change.

These pictures show the fraction 7/3 converted into an equivalent fraction 14/6.

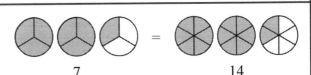

Now, 7/3 is a fraction, not a mixed number. You can see that from the picture because the whole pies have been split into fractional pieces. We consider it as seven thirds (slices).

$$\frac{7}{3} = \frac{14}{6}$$

Also, 7/3 is an **improper fraction** because its value is 1 or more. (Of course, 14/6 is also.)

A **proper fraction** is a fraction whose value is less than 1.

We use equivalent fractions also with mixed numbers and with improper fractions.

1. These are improper fractions. Split the slices in the right-hand picture. Write the equivalent fractions.

a. Split each slice into three.

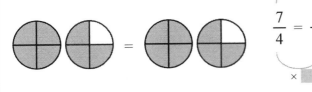

$$\frac{7}{4} = \frac{\quad}{\quad}$$

b. Split each slice in two.

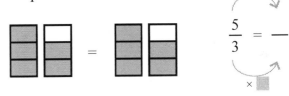

$$\frac{5}{3} = \frac{\quad}{\quad}$$

c. Split each slice in two.

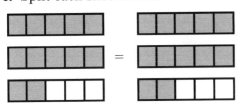

$$\frac{12}{5} = \frac{\quad}{\quad}$$

d. Split each slice into four.

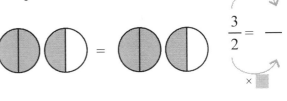

$$\frac{3}{2} = \frac{\quad}{\quad}$$

2. Fill in the missing numbers in these equivalent fractions and mixed numbers.

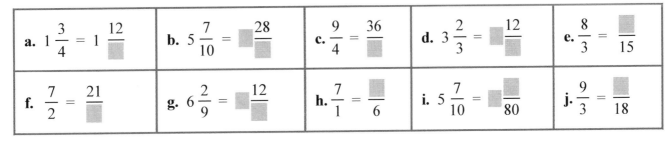

a. $1\frac{3}{4} = 1\frac{12}{\blacksquare}$	**b.** $5\frac{7}{10} = \blacksquare\frac{28}{\blacksquare}$	**c.** $\frac{9}{4} = \frac{36}{\blacksquare}$	**d.** $3\frac{2}{3} = \blacksquare\frac{12}{\blacksquare}$	**e.** $\frac{8}{3} = \frac{\blacksquare}{15}$
f. $\frac{7}{2} = \frac{21}{\blacksquare}$	**g.** $6\frac{2}{9} = \blacksquare\frac{12}{\blacksquare}$	**h.** $\frac{7}{1} = \frac{\blacksquare}{6}$	**i.** $5\frac{7}{10} = \blacksquare\frac{\blacksquare}{80}$	**j.** $\frac{9}{3} = \frac{\blacksquare}{18}$

3. Write the number 3 as a fraction using...

whole pies	halves	thirds	fourths	fifths	tenths	hundredths
$\dfrac{3}{1}$	$\dfrac{\blacksquare}{2}$					

4. Write the number 2 1/2 as a fraction using...

halves	fourths	sixths	eighths	tenths	twentieths	hundredths
$\dfrac{\blacksquare}{2}$						

5. If you can find an equivalent fraction, then write it. If you cannot find one, cross out the whole problem.

a. $\dfrac{5}{7} = \dfrac{}{28}$ The pieces were split into ____ .	**b.** $\dfrac{2}{5} = \dfrac{}{18}$ The pieces were split into ____ .	**c.** $\dfrac{1}{4} = \dfrac{}{14}$ The pieces were split into ____ .	**d.** $\dfrac{2}{3} = \dfrac{}{12}$ The pieces were split into ____ .	**e.** $\dfrac{5}{6} = \dfrac{8}{}$ The pieces were split into ____ .
f. $\dfrac{1}{6} = \dfrac{}{28}$ The pieces were split into ____ .	**g.** $\dfrac{2}{9} = \dfrac{}{63}$ The pieces were split into ____ .	**h.** $\dfrac{5}{4} = \dfrac{}{32}$ The pieces were split into ____ .	**i.** $\dfrac{1}{3} = \dfrac{5}{}$ The pieces were split into ____ .	**j.** $\dfrac{3}{8} = \dfrac{8}{}$ The pieces were split into ____ .

6. Explain in your own words when a problem of equivalent fractions is *not possible* to do. Use an example problem or problems in your explanation.

7. Make chains of equivalent fractions. Pay attention to the *patterns* in the numerators and in the denominators.

a. $\dfrac{3}{4} = \dfrac{\blacksquare}{8} = \quad = \quad = \quad = \quad = \quad = \quad$

b. $\dfrac{5}{3} = \dfrac{10}{6} = \dfrac{\blacksquare}{9} = \quad = \quad = \quad = \quad = \quad = \quad$

Adding and Subtracting Unlike Fractions

Cover the page below the black line. Then try to figure out the addition problems below.

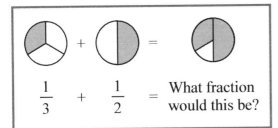

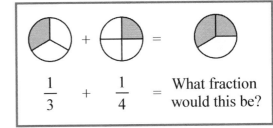

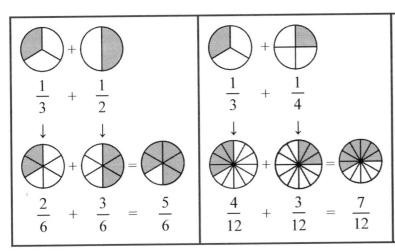

Did you solve the problems above?

The solution is this:

We convert the fractions so that they become *like* fractions (with a same denominator), using equivalent fractions.

Then we can add or subtract.

1. Write the fractions shown by the pie images. Convert them into equivalent fractions with the same denominator (like fractions), and then add them. Color the missing parts.

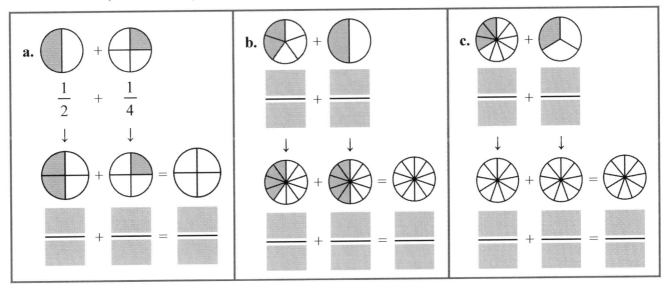

2. Convert the fractions to like fractions first, then add or subtract. In the bottom problems (d-f), you need to figure out what kind of pieces to use, but the *top* problems (a-c) will help you do that!

a.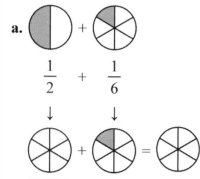

$$\frac{1}{2} + \frac{1}{6}$$

$$\downarrow \qquad \downarrow$$

$$\frac{}{} + \frac{1}{6} = \frac{}{}$$

b.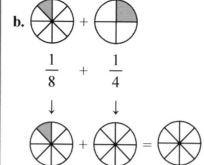

$$\frac{1}{8} + \frac{1}{4}$$

$$\downarrow \qquad \downarrow$$

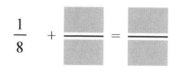

$$\frac{1}{8} + \frac{}{} = \frac{}{}$$

c.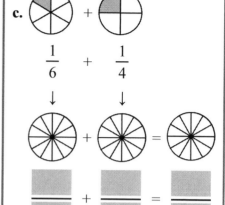

$$\frac{1}{6} + \frac{1}{4}$$

$$\downarrow \qquad \downarrow$$

$$\frac{}{} + \frac{}{} = \frac{}{}$$

d.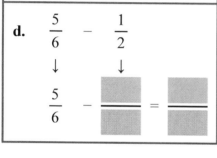

$$\frac{5}{6} - \frac{1}{2}$$

$$\downarrow \qquad \downarrow$$

$$\frac{5}{6} - \frac{}{} = \frac{}{}$$

e.

$$\frac{5}{8} - \frac{1}{4}$$

$$\downarrow \qquad \downarrow$$

$$\frac{}{} - \frac{}{} = \frac{}{}$$

f.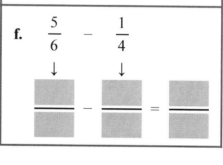

$$\frac{5}{6} - \frac{1}{4}$$

$$\downarrow \qquad \downarrow$$

$$\frac{}{} - \frac{}{} = \frac{}{}$$

g.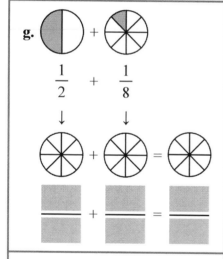

$$\frac{1}{2} + \frac{1}{8}$$

$$\downarrow \qquad \downarrow$$

$$\frac{}{} + \frac{}{} = \frac{}{}$$

h.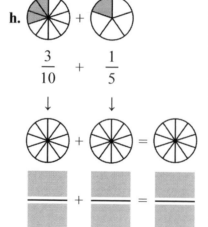

$$\frac{3}{10} + \frac{1}{5}$$

$$\downarrow \qquad \downarrow$$

$$\frac{}{} + \frac{}{} = \frac{}{}$$

i.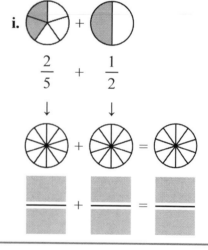

$$\frac{2}{5} + \frac{1}{2}$$

$$\downarrow \qquad \downarrow$$

$$\frac{}{} + \frac{}{} = \frac{}{}$$

j.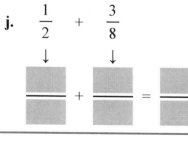

$$\frac{1}{2} + \frac{3}{8}$$

$$\downarrow \qquad \downarrow$$

$$\frac{}{} + \frac{}{} = \frac{}{}$$

k.

$$\frac{9}{10} - \frac{2}{5}$$

$$\downarrow \qquad \downarrow$$

$$\frac{}{} - \frac{}{} = \frac{}{}$$

l.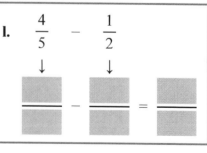

$$\frac{4}{5} - \frac{1}{2}$$

$$\downarrow \qquad \downarrow$$

$$\frac{}{} - \frac{}{} = \frac{}{}$$

3. Split the parts only in the *first* fraction so that both fractions will have the same kind of parts. Add.

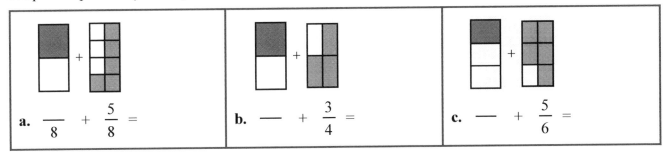

a. $\dfrac{}{8} + \dfrac{5}{8} =$

b. $\dfrac{}{} + \dfrac{3}{4} =$

c. $\dfrac{}{} + \dfrac{5}{6} =$

Now split the parts in *both* fractions so that they will have the same kind of parts. Add.

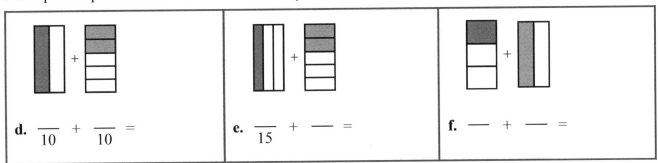

d. $\dfrac{}{10} + \dfrac{}{10} =$

e. $\dfrac{}{15} + \dfrac{}{} =$

f. $\dfrac{}{} + \dfrac{}{} =$

4. Fill in the table based on the problems above. What kind of parts did the two fractions have at first? What kind of parts did you use in the final addition?

Types of parts:	Converted to:	Types of parts:	Converted to:
a. 2nd parts and 8th parts	_8th_ parts	**d.** 2nd parts and 5th parts	_____ parts
b. 2nd parts and 4th parts	_____ parts	**e.** 3rd parts and 5th parts	_____ parts
c. 3rd parts and 6th parts	_____ parts	**f.** 3rd parts and 2nd parts	_____ parts

5. Now think: How can you know into what kind of parts to convert the fractions that you are adding? Can you see any patterns or rules in the table above?

6. Challenge: If you think you know what kind of parts to convert these fractions into, then try these problems. Do not worry if you don't know how to do them—we will study this in the next lesson.

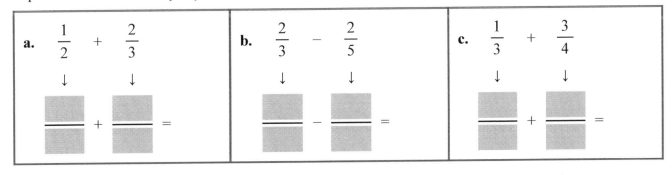

a. $\dfrac{1}{2} + \dfrac{2}{3}$

b. $\dfrac{2}{3} - \dfrac{2}{5}$

c. $\dfrac{1}{3} + \dfrac{3}{4}$

Finding the (Least) Common Denominator

Before adding or subtracting unlike fractions, first convert them into *like* fractions.

Before the conversion, we need to decide into what kinds of parts to convert them—in other words, what will be the new denominator for both. We call this denominator **the common denominator** because all of the converted fractions will have this same denominator in common.

To do the actual conversion, use the **principles you have learned concerning equivalent fractions**.

Example 1. $\dfrac{1}{6} + \dfrac{5}{9}$

$\downarrow \qquad \downarrow$

$\dfrac{3}{18} + \dfrac{10}{18} = \dfrac{13}{18}$

Notice that we use 18 as the common denominator. Why 18? You will find out soon, on the next page.

For now, notice that 1/6 is converted into 3/18 and 5/9 is converted into 10/18 using the rule for writing equivalent fractions. See the sidebar on the right if you've forgotten that. →

$$\overset{\times 3}{\frown} \qquad \overset{\times 2}{\frown}$$
$$\frac{1}{6} = \frac{3}{18} \qquad \frac{5}{9} = \frac{10}{18}$$
$$\underset{\times 3}{\smile} \qquad \underset{\times 2}{\smile}$$

1. You are given the common denominator. Convert the fractions using the rule for equivalent fractions. Then add or subtract. Note: sometimes you need to convert only one fraction, not both.

a. $\dfrac{1}{3} + \dfrac{3}{5}$

$\downarrow \qquad \downarrow$

$\dfrac{}{15} + \dfrac{}{15} =$

b. $\dfrac{6}{7} - \dfrac{1}{2}$

$\downarrow \qquad \downarrow$

$\dfrac{}{14} - \dfrac{}{14} =$

c. $\dfrac{1}{6} + \dfrac{2}{5}$

$\downarrow \qquad \downarrow$

$\dfrac{}{30} + \dfrac{}{30} =$

d. $\dfrac{5}{9} - \dfrac{1}{3}$

$\downarrow \qquad \downarrow$

$\dfrac{}{9} - \dfrac{}{9} =$

e. $\dfrac{1}{8} + \dfrac{3}{4}$

$\downarrow \qquad \downarrow$

$\dfrac{}{8} + \dfrac{}{8} =$

f. $\dfrac{5}{7} - \dfrac{2}{3}$

$\downarrow \qquad \downarrow$

$\dfrac{}{21} - \dfrac{}{21} =$

g. $\dfrac{2}{5} + \dfrac{1}{4}$

$\downarrow \qquad \downarrow$

$\dfrac{}{20} + \dfrac{}{20} =$

h. $\dfrac{5}{6} - \dfrac{3}{4}$

$\downarrow \qquad \downarrow$

$\dfrac{}{12} - \dfrac{}{12} =$

i. $\dfrac{3}{4} - \dfrac{3}{7}$

$\downarrow \qquad \downarrow$

$\dfrac{}{28} - \dfrac{}{28} =$

The common denominator has to be a <u>multiple</u> of each of the denominators.

In other words, the common denominator has to be in the multiplication table of the individual denominators. This also means that the common denominator has to be divisible by the individual denominators and the individual denominators have to "go into" the common denominator. See examples below.

Examples:

$\dfrac{2}{3} + \dfrac{1}{5} = \dfrac{\;\;\;}{15} + \dfrac{\;\;\;}{15}$ The common denominator must be a multiple of 5 and also a multiple of 3. Fifteen will work: it is in the multiplication table of 5 and also of 3.

$\dfrac{3}{8} - \dfrac{1}{6} = \dfrac{\;\;\;}{24} - \dfrac{\;\;\;}{24}$ Check the multiples of 8 (the skip-counting list): 0, 8, 16, 24, 32, *etc.* Compare to the multiples of 6: 0, 6, 12, 18, 24, 30, *etc.* We notice that **24** is the smallest number that is in both lists.

$\dfrac{7}{8} + \dfrac{3}{4} = \dfrac{7}{8} + \dfrac{\;\;\;}{8}$ We need a number that 4 can "go into" and that 8 can "go into." Actually, the smallest such number is 8 itself. So in this case, the 7/8 does not need to be converted at all; you just convert the 3/4 into 6/8.

2. Find a common denominator (c.d.) that will work with these fractions.

fractions to add/subtract	c.d.
a. 4th parts and 5th parts	
b. 3rd parts and 7th parts	
c. 10th parts and 2nd parts	

fractions to add/subtract	c.d.
d. 4th parts and 12th parts	
e. 2nd parts and 7th parts	
f. 9th parts and 6th parts	

3. Let's add and subtract. Use the common denominators you found above.

a. $\dfrac{4}{5} + \dfrac{1}{4}$
$\downarrow \qquad \downarrow$
$\dfrac{\;\;\;}{20} + \dfrac{\;\;\;}{20} =$

b. $\dfrac{2}{3} - \dfrac{1}{7}$
$\downarrow \qquad \downarrow$
$\dfrac{\;\;\;}{\;\;\;} - \dfrac{\;\;\;}{\;\;\;} =$

c. $\dfrac{3}{10} + \dfrac{1}{2}$
$\downarrow \qquad \downarrow$
$\dfrac{\;\;\;}{\;\;\;} + \dfrac{\;\;\;}{\;\;\;} =$

d. $\dfrac{4}{12} + \dfrac{1}{4}$
$\downarrow \qquad \downarrow$
$\dfrac{\;\;\;}{\;\;\;} + \dfrac{\;\;\;}{\;\;\;} =$

e. $\dfrac{1}{2} - \dfrac{2}{7}$
$\downarrow \qquad \downarrow$
$\dfrac{\;\;\;}{\;\;\;} - \dfrac{\;\;\;}{\;\;\;} =$

f. $\dfrac{5}{6} - \dfrac{4}{9}$
$\downarrow \qquad \downarrow$
$\dfrac{\;\;\;}{\;\;\;} - \dfrac{\;\;\;}{\;\;\;} =$

<table>
<tr><td colspan="2">You can always multiply the denominators to get a common denominator. However, you can often find a smaller number than the denominator you get by multiplying the denominators.</td></tr>
<tr>
<td>$\dfrac{7}{10}$ and $\dfrac{1}{15}$</td>
<td>You could use $10 \times 15 = 150$, but let's look at the lists of multiples:

Multiples of 10:   0, 10, 20, <u>30</u>, 40, 50, ...
Multiples of 15:   0, 15, <u>30</u>, 45, 60, 75 ...

So, 30 works as well, and it is smaller!
It is the least common denominator.</td>
</tr>
<tr>
<td>$\dfrac{2}{7}$ and $\dfrac{1}{6}$</td>
<td>One possibility is $7 \times 6 = 42$, but let's check the multiples of 6 to make sure:

Multiples of 6:   0, 6, 12, 18, 24, 30, 36, 42, 48, ...
None of those are in the multiplication table of 7, except 42.

So, 42 is the Least Common Denominator (LCD).</td>
</tr>
</table>

4. Find the least common denominator (LCD) for adding or subtracting these fractions. You may use the space for writing out lists of multiples.

fractions	LCD
a. $\dfrac{5}{12}$ + $\dfrac{3}{8}$ $\downarrow$ $\downarrow$ + =	
b. $\dfrac{7}{4}$ − $\dfrac{9}{11}$ $\downarrow$ $\downarrow$ − =	
c. $\dfrac{1}{12}$ + $\dfrac{1}{9}$ $\downarrow$ $\downarrow$ + =	
d. $\dfrac{7}{8}$ − $\dfrac{4}{9}$ $\downarrow$ $\downarrow$ − =	

Add and Subtract: More Practice

Example 1. Amy added: $\frac{2}{5} + \frac{2}{7} = \frac{4}{35}$, but that is not right! How can you tell? The answer, 4/35, is a *very small number*. It is much less than 2/5! In fact, 4/35 is close to 5/35 = 1/7.	The correct answer is: $\frac{2}{5} + \frac{2}{7}$ $\downarrow \quad \downarrow$ $\frac{14}{35} + \frac{10}{35} = \frac{24}{35}$
Example 2. Sam added: $\frac{5}{6} + \frac{3}{4} = \frac{8}{10}$, but it is wrong! How can you tell? The answer, 8/10, is less than 1, whereas the sum of 5/6 and 3/4 is surely more than 1.	The correct answer is: $\frac{5}{6} + \frac{3}{4}$ $\downarrow \quad \downarrow$ $\frac{10}{12} + \frac{9}{12} = \frac{19}{12} = 1\frac{7}{12}$

Always check if your answer is reasonable.

Once you master adding and subtracting unlike fractions, you will have learned the *hardest part* of fraction math! **Congratulations!** Multiplication and division of fractions are actually easier.

1. Use your "fraction number sense" to explain why these answers are wrong. Find the correct answer.

a. Amanda added: $\frac{5}{8} + \frac{5}{8} = \frac{10}{16}$ How can you tell it is wrong? Correct answer:	**b.** Robert subtracted: $\frac{7}{9} - \frac{1}{2} = \frac{6}{7}$ How can you tell it is wrong? Correct answer:

2. Explain why Olivia's answer must be wrong. Then find the correct answer.

Olivia subtracted: $\frac{7}{12} - \frac{1}{4} = \frac{1}{2}$ Correct answer:

How can you tell it is wrong?

3. Write the letters that match the answers in the boxes to solve the riddle. Check that your answers are reasonable.

Why did the banana go to the doctor?

$\frac{23}{36}$ $\frac{5}{6}$ $\frac{7}{15}$ $\frac{3}{10}$ $\frac{1}{10}$ $\frac{4}{15}$ $\frac{5}{6}$ $\frac{23}{30}$ $\frac{7}{45}$ $\frac{8}{15}$ $\frac{31}{30}$ $\frac{17}{28}$

☐ ☐ ☐ ☐ ☐ ☐ ☐   ☐ ☐   ☐ ☐ ☐

$\frac{17}{21}$ $\frac{9}{10}$ $\frac{9}{35}$ $\frac{83}{72}$ $\frac{11}{24}$ $\frac{11}{24}$ $\frac{27}{40}$ $\frac{23}{30}$ $\frac{5}{9}$ $\frac{1}{6}$ $\frac{1}{14}$ $\frac{104}{63}$ $\frac{7}{6}$ $\frac{13}{20}$

☐ ☐ ☐   ☐ ☐ ☐ ☐ ☐ ☐ ☐   ☐ ☐ ☐ ☐ !

L $\frac{1}{2} + \frac{2}{3}$	**W** $\frac{1}{5} + \frac{1}{3}$	**S** $\frac{2}{5} - \frac{2}{15}$
$\frac{}{6} + \frac{}{6} =$		
E $\frac{2}{6} + \frac{1}{2}$	**C** $\frac{2}{3} - \frac{1}{5}$	**G** $\frac{5}{6} - \frac{2}{3}$
A $\frac{1}{10} + \frac{1}{5}$	**I** $\frac{1}{3} + \frac{13}{30}$	**U** $\frac{3}{5} - \frac{1}{2}$

76

These problems belong to the riddle on the previous page, as well.

E $\dfrac{1}{3} + \dfrac{1}{8}$	T $\dfrac{5}{9} - \dfrac{2}{5}$	N $\dfrac{2}{3} + \dfrac{1}{7}$
E $\dfrac{5}{6} - \dfrac{3}{8}$	O $\dfrac{1}{2} + \dfrac{4}{10}$	W $\dfrac{4}{7} - \dfrac{1}{2}$
L $\dfrac{4}{5} - \dfrac{3}{20}$	S $\dfrac{6}{7} - \dfrac{1}{4}$	B $\dfrac{2}{9} + \dfrac{5}{12}$
T $\dfrac{6}{7} - \dfrac{3}{5}$	A $\dfrac{1}{5} + \dfrac{5}{6}$	P $\dfrac{11}{8} - \dfrac{2}{9}$
N $\dfrac{2}{3} - \dfrac{1}{9}$	E $\dfrac{10}{7} + \dfrac{2}{9}$	L $\dfrac{7}{8} - \dfrac{1}{5}$

Puzzle Corner

Find the fractions that can go into the puzzles.

Hint: If the answer has a denominator of 15, think what the denominators of the two fractions could have been.

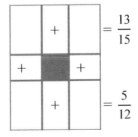

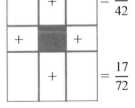

Adding and Subtracting Mixed Numbers

Adding and subtracting mixed numbers with unlike fractional parts is not difficult:

1. First convert the unlike fractional parts into like fractions.
2. Then add or subtract the mixed numbers.

You already know how to do both steps. Study the example below.

$$2 \frac{1}{2} + 1 \frac{7}{8} = 2 \frac{4}{8} + 1 \frac{7}{8} = 3 \frac{11}{8} = 4 \frac{3}{8}$$

Convert 1/2 into 4/8. Then add. Lastly, change the mixed number 3 11/8 into 4 3/8.

$$2 \frac{1}{2} \Rightarrow 2 \frac{4}{8}$$
$$+ 1 \frac{7}{8} \qquad + 1 \frac{7}{8}$$
$$\overline{\qquad} \qquad \overline{\qquad}$$
$$3 \frac{11}{8} \Rightarrow 4 \frac{3}{8}$$

1. Split the pieces so that you have *like* fractional parts. Write an addition sentence.

a.

b.

c.

2. First convert the fractional parts into like fractions, then add.

a. $6 \frac{2}{3} \Rightarrow 6 \frac{}{15}$

$+ 3 \frac{1}{5} \qquad + 3 \frac{}{}$
$\overline{\qquad} \qquad \overline{\qquad}$

b. $10 \frac{1}{8} \Rightarrow$

$+ 3 \frac{2}{5}$
$\overline{\qquad} \qquad \overline{\qquad}$

c. $17 \frac{1}{16} \Rightarrow$

$+ 3 \frac{3}{8}$
$\overline{\qquad} \qquad \overline{\qquad}$

Split the piece so you have sixth parts.

$$2\frac{1}{2} - 1\frac{2}{3} = 2\frac{3}{6} - 1\frac{4}{6}$$

$$\downarrow$$

Now, rename 2 3/6 as 1 9/6. This is the same process as regrouping.

$$= 1\frac{9}{6} - 1\frac{4}{6} = \frac{5}{6}$$

$$2\frac{1}{2} \Rightarrow 2\frac{\cancel{3}}{6} \quad 1\frac{9}{6}$$
$$-1\frac{2}{3} \qquad -1\frac{4}{6}$$
$$\overline{} \qquad \overline{\frac{5}{6}}$$

3. Split the pieces in such a way that you can cross out what is indicated. Write a subtraction sentence.

a.  Cross out $1\frac{3}{8}$.

b. Cross out $1\frac{1}{3}$.

c. Cross out $1\frac{7}{10}$.

d. Cross out $1\frac{4}{9}$.

4. First convert the fractional parts into like fractions, then subtract. You may need to regroup.

a. $5\frac{1}{2}$ $\Rightarrow$

$- 2\frac{4}{5}$

b. $15\frac{4}{8}$ $\Rightarrow$

$- 8\frac{5}{6}$

c. $16\frac{5}{9}$ $\Rightarrow$

$- 10\frac{1}{2}$

d. $4\frac{1}{6}$ $\Rightarrow$

$- 2\frac{3}{5}$

e. $11\frac{1}{12}$ $\Rightarrow$

$- 3\frac{1}{4}$

f. $8\frac{2}{9}$ $\Rightarrow$

$- 2\frac{3}{4}$

5. First convert the fractional parts into like fractions, then add. Lastly, change your final answer so that the fractional part is not more than 1 whole.

a. $4 \dfrac{1}{2} \Rightarrow$ $4 \dfrac{}{10}$

$+\ 3 \dfrac{4}{5}$ $3 \dfrac{}{10}$

_____   _____ $\Rightarrow$

b. $5 \dfrac{5}{6} \Rightarrow$

$+\ 7 \dfrac{2}{3}$

_____   _____ $\Rightarrow$

c. $3 \dfrac{5}{6} \Rightarrow$

$+\ 2 \dfrac{7}{8}$ $+$

_____   _____ $\Rightarrow$

d. $9 \dfrac{5}{7} \Rightarrow$

$+\ 7 \dfrac{2}{3}$

_____   _____ $\Rightarrow$

6. First convert the fractional parts into like fractions. Then add or subtract.

a. $5 \dfrac{3}{4} - 1 \dfrac{7}{8}$

b. $8 \dfrac{9}{15} + 5 \dfrac{4}{5}$

c. $3 \dfrac{2}{9} - 1 \dfrac{1}{3}$

d. $7 \dfrac{2}{7} - 2 \dfrac{1}{2}$

e. $8 \dfrac{3}{10} + 2 \dfrac{4}{5}$

f. $6 \dfrac{2}{3} - 1 \dfrac{1}{7}$

Example 3. A recipe calls for 1 1/3 cups of flour and 1/2 cup of coconut flakes. How many total cups are these two dry ingredients?

Harry added: $1\frac{1}{3} + \frac{1}{2} = 1\frac{2}{5}$, and gave the answer 1 2/5 cups of dry ingredients.

That is wrong, and you can *easily* see that, because <u>2/5 is less than 1/2</u>! You cannot add 1/3 and 1/2 and get an answer that is less than 1/2.

Always check if your answer is reasonable.

---

7. Henry's two heaviest school books weigh 1 3/4 lb and 1 11/16 lb.

   a. What is their total weight in *pounds*?

   b. Remember that 1 lb = 16 oz. Now change the total weight into pounds and ounces.

   (**c.** Do the same with your heaviest school books.)

---

8. Sally needs 1 1/4 meters of material to make a blouse and 8/10 of a meter to make a skirt.

   a. Add the mixed numbers to find how many meters of material she needs for both of them.

   b. Now solve the problem another way, by changing both measurements into centimeters first.

---

9. A company divided a project so that Mark would do 1/10 of it, Leslie would do 1/2 of it, and Jerry the rest. What part (fraction) of the job was left for Jerry?

10. Cindy needs to make two cakes, one batch of pancakes, and some sauce.
    She needs 3 1/2 dl (*deciliters*) of flour for a cake, 5 dl of flour for a batch of pancakes,
    and 3/4 dl of flour for the sauce.

    A 1 kg bag of flour is about 15 dl.
    Will one bag of flour be enough for her three recipes?

11. Bob paid 19/100 of his salary in taxes and 2/10 of it as a loan payment.

    a. What part (what fraction) of his salary is left after those?

    b. Calculate how many dollars Bob has left if his initial salary was $2,000.

12. Mom made a full pitcher of smoothie for the family. She gave 1/8 of it to Anna,
    1/8 of it to Jack, 1/4 of it to dad, and drank 1/4 of it herself.

    a. What fraction of the full pitcher is left now?

    b. If the original amount of smoothie was 2 liters,
    how many liters of it is left now?

13. Lily's notebook is 3 1/4 in wide and 6 1/8 in long.
    She wants to glue a picture on the front so that
    the margins on all sides are 3/4 in.

    What size should her picture be (width and length)?

    picture

Comparing Fractions

Sometimes it is easy to know which fraction is the greater of the two. Study the examples below!

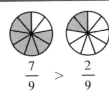

$$\frac{7}{9} > \frac{2}{9}$$

With **like fractions**, all you need to do is to check **which fraction has more "slices,"** and that fraction is greater.

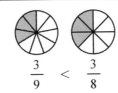

$$\frac{3}{9} < \frac{3}{8}$$

If both fractions have the **same number of pieces**, then the one with bigger pieces is greater.

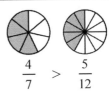

$$\frac{4}{7} > \frac{5}{12}$$

Sometimes you can **compare to 1/2**. Here, 4/7 is clearly more than 1/2, and 5/12 is clearly less than 1/2.

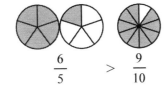

$$\frac{6}{5} > \frac{9}{10}$$

Any fraction that is bigger than one must also be bigger than any fraction that is less than one. Here, 6/5 is more than 1, and 9/10 is less than 1.

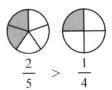

$$\frac{2}{5} > \frac{1}{4}$$

If you can imagine the pie pictures in your mind, you can sometimes "see" which fraction is bigger. For example, it is easy to see that 2/5 is more than 1/4.

1. Compare the fractions, and write > , < , or = .

a. $\frac{1}{8}$ $\frac{1}{10}$		**b.** $\frac{4}{9}$ $\frac{1}{2}$		**c.** $\frac{6}{10}$ $\frac{1}{2}$		**d.** $\frac{3}{9}$ $\frac{3}{7}$	
e. $\frac{8}{11}$ $\frac{4}{11}$		**f.** $\frac{7}{4}$ $\frac{7}{6}$		**g.** $\frac{5}{14}$ $\frac{5}{9}$		**h.** $\frac{4}{20}$ $\frac{2}{20}$	
i. $\frac{2}{11}$ $\frac{2}{5}$		**j.** $\frac{1}{2}$ $\frac{5}{8}$		**k.** $\frac{3}{6}$ $\frac{1}{2}$		**l.** $\frac{1}{20}$ $\frac{1}{8}$	
m. $\frac{1}{2}$ $\frac{3}{4}$		**n.** $\frac{8}{7}$ $\frac{3}{3}$		**o.** $\frac{49}{100}$ $\frac{61}{100}$		**p.** $\frac{7}{8}$ $\frac{8}{7}$	
q. $\frac{9}{10}$ $\frac{3}{4}$		**r.** $\frac{6}{5}$ $\frac{3}{4}$		**s.** $\frac{4}{4}$ $\frac{9}{11}$		**t.** $\frac{1}{3}$ $\frac{3}{9}$	

Sometimes none of the "tricks" explained in the previous page work, but we do have one more up our sleeve!

Convert both fractions into like fractions. Then compare.

In the picture on the right, it is hard to be sure if 3/5 is really more than 5/9. Convert both into 45th parts, and then it is easy to see that 27/45 is more than 25/45. Not by much, though!

$$\frac{3}{5} \qquad \frac{5}{9}$$

$$\downarrow \qquad \downarrow$$

$$\frac{27}{45} > \frac{25}{45}$$

2. Convert the fractions into like fractions, and then compare them.

a. $\dfrac{2}{3}$ $\dfrac{5}{8}$ $\downarrow$ $\downarrow$	**b.** $\dfrac{5}{6}$ $\dfrac{7}{8}$ $\downarrow$ $\downarrow$	**c.** $\dfrac{1}{3}$ $\dfrac{3}{10}$ $\downarrow$ $\downarrow$	**d.** $\dfrac{8}{12}$ $\dfrac{7}{10}$ $\downarrow$ $\downarrow$
e. $\dfrac{5}{8}$ $\dfrac{7}{12}$ $\downarrow$ $\downarrow$	**f.** $\dfrac{11}{8}$ $\dfrac{14}{10}$ $\downarrow$ $\downarrow$	**g.** $\dfrac{6}{10}$ $\dfrac{58}{100}$ $\downarrow$ $\downarrow$	**h.** $\dfrac{6}{5}$ $\dfrac{11}{9}$ $\downarrow$ $\downarrow$
i. $\dfrac{7}{10}$ $\dfrac{5}{7}$ $\downarrow$ $\downarrow$	**j.** $\dfrac{43}{100}$ $\dfrac{3}{10}$ $\downarrow$ $\downarrow$	**k.** $\dfrac{9}{8}$ $\dfrac{8}{7}$ $\downarrow$ $\downarrow$	**l.** $\dfrac{7}{10}$ $\dfrac{2}{3}$ $\downarrow$ $\downarrow$

3. One cookie recipe calls for 1/2 cup of sugar. Another one calls for 2/3 cup of sugar. Which uses more sugar, a triple batch of the first recipe, or a double batch of the second?

How much more?

4. Compare the fractions using any method.

a. $\dfrac{5}{12}$ $\dfrac{3}{8}$	**b.** $\dfrac{5}{12}$ $\dfrac{4}{11}$	**c.** $\dfrac{3}{10}$ $\dfrac{1}{5}$	**d.** $\dfrac{3}{8}$ $\dfrac{4}{7}$
e. $\dfrac{4}{15}$ $\dfrac{1}{3}$	**f.** $\dfrac{5}{6}$ $\dfrac{11}{16}$	**g.** $\dfrac{7}{6}$ $\dfrac{10}{8}$	**h.** $\dfrac{5}{12}$ $\dfrac{5}{8}$
i. $\dfrac{3}{4}$ $\dfrac{4}{11}$	**j.** $\dfrac{13}{10}$ $\dfrac{9}{8}$	**k.** $\dfrac{2}{13}$ $\dfrac{1}{5}$	**l.** $\dfrac{1}{10}$ $\dfrac{1}{11}$

5. A coat costs $40. Which is a bigger discount:
   1/4 off the normal price, or 3/10 off the normal price?

   Does your answer change if the original price
   of the coat was $60 instead? Why or why not?

6. Here are three number lines that are divided respectively into halves, thirds, and fifths. Use them to
   help you put the given fractions in order, from the least to the greatest.

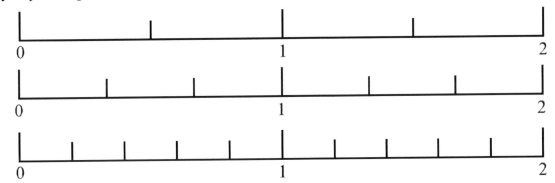

a. $\dfrac{1}{3}, \dfrac{2}{5}, \dfrac{2}{3}, \dfrac{1}{5}, \dfrac{1}{2}$ **b.** $\dfrac{7}{5}, \dfrac{3}{2}, \dfrac{4}{3}, \dfrac{6}{5}, \dfrac{2}{2}$

___ < ___ < ___ < ___ < ___   ___ < ___ < ___ < ___ < ___

7. Write the three fractions in order.

a. $\dfrac{7}{8}$, $\dfrac{9}{10}$, $\dfrac{7}{9}$	b. $\dfrac{1}{3}$, $\dfrac{4}{10}$, $\dfrac{2}{9}$
___ < ___ < ___	___ < ___ < ___

8. Rebecca made a survey of a group of 600 women. She found that 1/3 of them never exercised, that 22/100 of them swam regularly, 1/5 of them jogged regularly, and the rest were involved in other sports.

a. Which was a bigger group, the women who jogged or the women who swam?

b. What fraction of this group of women exercise?

c. *How many women* in this group exercise?

d. How many women in this group swim?

The seven dwarfs could not divide a pizza into seven equal slices. The oldest suggested, "Let's cut it into eight slices, let each dwarf have one piece, and give the last piece to the dog."

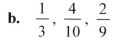
Puzzle Corner

Then another dwarf said, "No! Let's cut it into 12 slices instead, and give each of us 1 ½ of those pieces, and the dog gets the 1 ½ pieces left over."

Which suggestion would give more pizza to the dog?

Measuring in Inches

Here are four rulers that all measure in inches. They are *not* to scale. Instead, they are magnified to be bigger than real rulers, so you can see how they are divided into parts.

The tick-marks are:

every 1/2-inch:

every 1/4-inch:

every 1/8-inch:

every 1/16-inch:

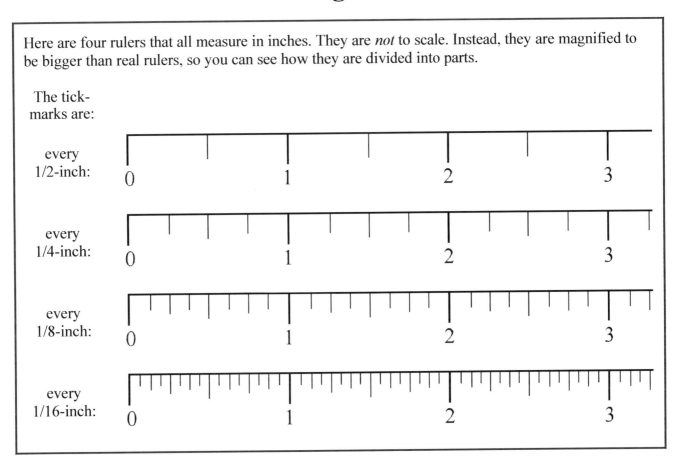

1. Find the ½-inch mark, 1 ½-inch mark, and 2 ½-inch mark on all of the rulers above.

2. Find the ¼-inch mark, the ¾-inch mark, the 1 ¼-inch mark, the 1 ¾-inch mark, the 2 ¼-inch mark, the 2 ¾-inch mark, and the 3 ¼-inch mark on the bottom three rulers above.

3. On the ruler that measures in 8th parts of an inch, find and label tick marks for these points: the 1/8-inch point, the 5/8-inch point, the 7/8-inch point, the 1 5/8-inch point, and the 2 3/8-inch point.

   Also, find these *same* points on the ruler that measures in 16th parts of an inch.

4. Look at the ruler that measures in 16th parts of an inch. On that ruler find tick marks for these points:

   - 3/16 inch
   - 7/16 inch
   - 11/16 inch

   - 1 1/8 inches
   - 2 3/8 inches
   - 7/8 inch

   - 1/4 inch
   - 1 1/4 inches
   - 2 3/4 inches

5. Measure the following colored lines with the rulers given. If the end of the line does not fall exactly on a tick mark, then read the mark that is CLOSEST to the end of the line.

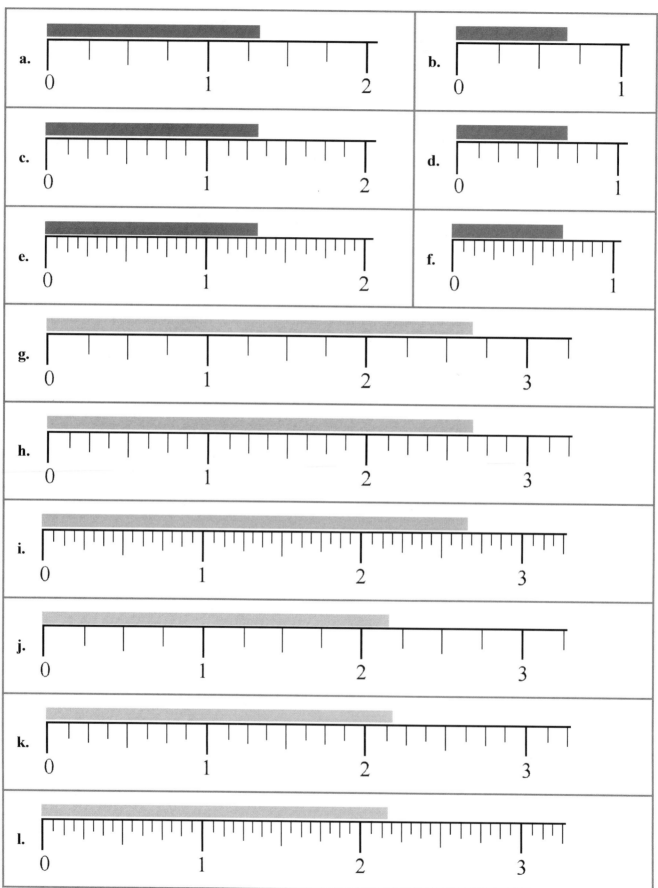

6. Measure the following lines using different rulers. Cut out the rulers from the bottom of this page.

a. ▬▬▬▬▬▬▬▬▬▬▬▬▬▬▬

Using the 1/4-inch ruler: _____ in.

Using the 1/8-inch ruler: _____ in.

Using the 1/16-inch ruler: _____ in.

b. ▬▬▬▬

Using the 1/4-inch ruler: _____ in.

Using the 1/8-inch ruler: _____ in.

Using the 1/16-inch ruler: _____ in.

c. ▬▬▬▬▬▬▬▬▬▬▬▬

Using the 1/4-inch ruler: _____ in.

Using the 1/8-inch ruler: _____ in.

Using the 1/16-inch ruler: _____ in.

d. ▬▬▬▬▬

Using the 1/4-inch ruler: _____ in.

Using the 1/8-inch ruler: _____ in.

Using the 1/16-inch ruler: _____ in.

e. ▬▬▬▬

Using the 1/4-inch ruler: _____ in.

Using the 1/8-inch ruler: _____ in.

Using the 1/16-inch ruler: _____ in.

f. ▬▬▬▬▬▬▬▬▬

Using the 1/4-inch ruler: _____ in.

Using the 1/8-inch ruler: _____ in.

Using the 1/16-inch ruler: _____ in.

You may cut out the following rulers:

1   2   3   4   5

1   2   3   4   5

1   2   3   4   5

7. Find six items in your home that you can measure with your ruler and measure them.

    a. _____ _____ in.   **b.** _____ _____ in.

    c. _____ _____ in.   **d.** _____ _____ in.

    e. _____ _____ in.   **f.** _____ _____ in.

8. Carefully measure the sides of the quadrilateral at the right, and find its perimeter.

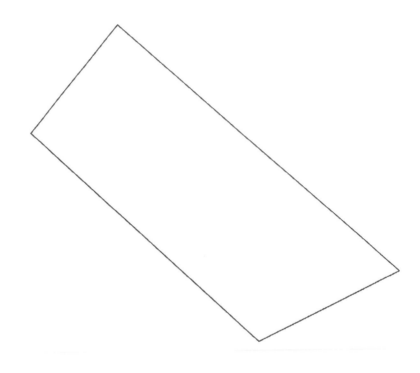

9. A small rectangular bulletin board measures 15 3/4 in by 9 1/8 in. What is its perimeter?

10. Janet checked the amount of sugar in 10 different cookie recipes. The amounts, in cups, were:

    1 1/2   1 3/8   1   1 3/4   1 1/2   1 1/8   1 1/4   1 1/4   1 1/2   3/4

    a. Make a line plot from this data (below) by drawing an X-mark for each measurement above the number line.

    b. If Janet made the recipe with the least amount of sugar three times, how much sugar would she need?

    c. If Janet made the recipe with the largest amount of sugar three times, how much sugar would she need?

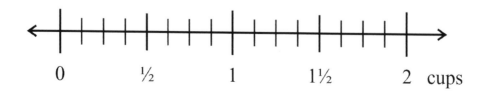

90

11. Make a line plot from these measurements (lengths of cockroaches, in inches, in Jake's collection):

1 1/4 1 1/8 1 1/8 1 1/2 1 1 1/8 1 3/8 1 3/4 1 3/8 7/8 1 1/4 2 1/8 1/2 1 1/4 1 1/4

1 1/2 1 1/2 1 1/2 1 5/8

This time, you will need to do the scaling on the number line.

b. What is the mode of this data set?

c. Jake took his five longest cockroaches, and placed them end-to-end. How long a "train" did they form?

12. Measure a bunch of pencils to the nearest 1/8 or 1/16 of an inch. Then make a line plot of your data.

Mixed Review

1. In what place is the underlined digit? What is its value?

a. 452,9<u>1</u>2,980	b. <u>6</u>,219,455,221
Place: _____	Place: _____
Value: _____	Value: _____

2. How many seconds are there in an hour?

   How many seconds are there in a day?

3. The Hewitts family homeschools all but 12 weeks of the year, five days a week, about five hours a day. How many hours of school do they have in a year?

4. Solve by multiplying in columns. Estimate first.

   2.11×6.8

   Estimate: _____

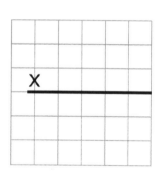

5. Multiply and divide.

a. $0.34 \div 10$ = _____	b. 100×0.098 = _____	c. $19 \div 10^3$ = _____
$2.1 \div 100$ = _____	$1,000 \times 46.7$ = _____	$10^4 \times 0.03$ = _____

6. Convert between the measuring units.

a. 5,070 g = _____ kg	b. 0.6 L = _____ ml	c. 0.06 km = _____ m
2.5 kg = _____ g	10,500 ml = _____ L	2,600 m = _____ km

7. First, multiply *both* the divisor and the dividend by 10, 100, or 1000 so that the divisor becomes a whole number. Then divide.

a. $82.50 \div 0.06$	**b.** $48.302 \div 0.2$

8. One cup of plain yogurt costs $2.40, a cup of strawberry yogurt costs $0.15 less than plain yogurt, and a cup of plum yogurt costs $0.30 more than plain yogurt. What is the total cost if you buy one cup of each kind of yogurt?

9. The price of Shirt A is $6.29. It is one-third of the price of Shirt B. Find the price of Shirt B.

10. Fill in Rachel's reasoning for solving $1{,}000 \times 0.007$.

Multiplying as if there was no decimal point, I get $1{,}000 \times$ _____ . That equals _____.

Then, since my answer has to have thousandths, it needs _____ decimal digits.

So, the final answer is _____.

11. **a.** Make a histogram from the data in the frequency table on the right.

Height in cm	Number of people
120...129	4
130...139	10
140...149	41
150...159	82

Height in cm	Number of people
160...169	95
170...179	61
180...189	39
190...199	6

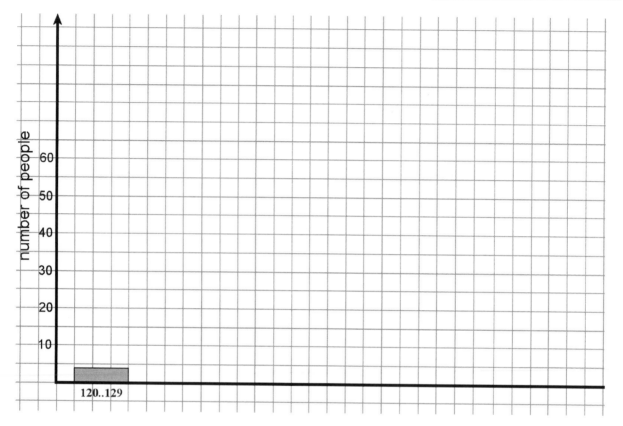

b. How many people were short (less than 140 cm tall)?

c. How many were tall (180 cm or taller)?

d. Most adults are 160 cm tall or taller. Use this fact to guess (estimate) how many children and how many adults were in this group.

e. Could this data come from...

- a group of elementary school children?

- a group of people who were at a swimming pool?

- a group of elderly women in an old people's home?

   Explain your reasoning.

94

Chapter 6 Review

1. Write as fractions. Think of the shortcut.

a. $9\frac{1}{2}$	**b.** $5\frac{6}{11}$	**c.** $8\frac{2}{7}$	**d.** $5\frac{6}{100}$

2. Write as mixed numbers.

a. $\frac{41}{10}$	**b.** $\frac{19}{3}$	**c.** $\frac{28}{9}$	**d.** $\frac{32}{12}$

3. For the division problem $23 \div 6 = 3$ R5, write a corresponding problem where a fraction is changed into a mixed number.

4. Subtract. Regroup if necessary. Check that your answer is reasonable.

a. $9\frac{4}{8}$ $-3\frac{7}{8}$	**b.** $12\frac{3}{20}$ $-5\frac{11}{20}$	**c.** $10\frac{3}{5}$ $-5\frac{1}{3}$

5. Add and subtract. Check that your answer is reasonable.

a. $\frac{5}{7} + \frac{1}{3}$	**b.** $\frac{3}{10} + \frac{1}{3}$
c. $3\frac{2}{7} - 1\frac{6}{7}$	**d.** $2\frac{4}{5} + 3\frac{1}{4}$

6. Compare the fractions, and write $<$, $>$, or $=$ in the box.

a. $\dfrac{1}{2}$ ☐ $\dfrac{3}{5}$	**b.** $\dfrac{3}{11}$ ☐ $\dfrac{1}{3}$	**c.** $\dfrac{7}{10}$ ☐ $\dfrac{70}{100}$	**d.** $\dfrac{1}{4}$ ☐ $\dfrac{28}{100}$
e. $\dfrac{2}{3}$ ☐ $\dfrac{8}{9}$	**f.** $\dfrac{1}{4}$ ☐ $\dfrac{2}{15}$	**g.** $\dfrac{21}{16}$ ☐ $\dfrac{25}{16}$	**h.** $\dfrac{5}{11}$ ☐ $\dfrac{1}{2}$

7. Betty uses 3 1/8 feet of material to make one shirt. She has one piece
   that is 5 1/2 feet and another piece that is 4 1/4 feet. She made one
   shirt from *each* piece of material.
   How much material does she have left now, in total?

8. Of a piece of land, 32/100 is planted in wheat, 42/100 is planted
   in barley, 2/10 is planted in oats, and the remainder is resting.
   What part (fraction) of the land is resting?

9. Which is a better deal: 1/5 off of a book that costs $35,
   or 2/11 off of a book that costs $33?

   Would the situation change if both deals involved a book
   that costs $50? Explain.

Chapter 7: Fractions: Multiply and Divide
Introduction

This is another long chapter devoted solely to fractions. It rounds out our study of fraction arithmetic. (If you feel that your student(s) would benefit from taking a break from fractions, you can optionally have them study chapter 8 on geometry in between chapters 6 and 7.)

We start out by simplifying fractions. Since this process is the opposite of making equivalent fractions, studied in chapter 6, it should be relatively simple for students to understand. We also use the same visual model, just backwards: This time the pie pieces are joined together instead of split apart.

Next comes multiplying a fraction by a whole number. Since this can be solved by repeated addition, it is not a difficult concept at all.

Multiplying a fraction by a fraction is first explained as taking a certain part of a fraction, in order to teach the concept. After that, students are shown the usual shortcut for the multiplication of fractions.

Simplifying before multiplying is a process that is not absolutely necessary for fifth graders. I have included it here because it prepares students for the same process in future algebra studies and because it makes fraction multiplication easier. I have also tried to include explanations of *why* we are allowed to simplify before multiplying. These explanations are actually *proofs*. I feel it is a great advantage for students to get used to mathematical reasoning and proof methods well before they start high school geometry.

Then, we find the area of a rectangle with fractional side lengths, and show that the area is the same as it would be found by multiplying the side lengths. Students multiply fractional side lengths to find areas of rectangles, and represent fraction products as rectangular areas.

Students also multiply mixed numbers, and study how multiplication can be seen as resizing or scaling. This means, for example, that the multiplication $(2/3) \times 18$ km can be thought of as finding two-thirds of 18 km.

Next, we study division of fractions in special cases. The first one is seeing fractions *as* divisions; in other words recognizing that 5/3 is the same as $5 \div 3$. This of course gives us a means of dividing whole numbers and getting fractional answers (for example, $20 \div 6 = 3\ 2/6$).

Then students encounter sharing divisions with fractions. For example, if two people share equally 4/5 of a pizza, how much will each person get? This is represented by the division $(4/5) \div 2 = 2/5$. Another case we study is dividing unit fractions by whole numbers (such as $(1/2) \div 4$).

We also divide whole numbers by unit fractions, such as $6 \div (1/3)$. Students will solve these thinking how many times the divisor "fits into" the dividend.

The last lesson is an introduction to ratios, and is optional. Ratios will be studied a lot in 6th and 7th grades, including in connection with proportions. We are laying the groundwork for that here.

The Lessons in Chapter 7

Helpful Resources on the Internet

Fraction Videos 2: Multiplication and Division
My own videos that cover multiplying and dividing fractions.
http://www.mathmammoth.com/videos/fractions_2.php

REDUCING/SIMPLIFYING FRACTIONS

Canceling Demonstration
Watch a movie that uses circles to demonstrate how to rename to lowest terms with canceling.
http://www.visualfractions.com/cancel/

Reduce Fractions Shoot
Reduce the fraction on the screen to the lowest terms by clicking the correct answer.
http://www.sheppardsoftware.com/mathgames/fractions/reduce_fractions_shoot.htm

Fraction Worksheets: Simplifying and Equivalent Fractions
Create custom-made worksheets for fraction simplification and equivalent fractions.
http://www.homeschoolmath.net/worksheets/fraction.php

Reducing Fractions to Lowest Terms
This is a simple online exercise that you can use for extra practice.
http://www.mathgames.com/skill/3.46-reducing-fractions-to-lowest-terms

Fractions Booster
How much pizza is left? Be sure to reduce the answer down to lowest terms!
http://www.tgfl.org.uk/tgfl/custom/resources_ftp/netmedia_ll/ks2/maths/fractions/level5.htm

Add and Subtract Fractions Game
Solve the given equation and gobble the correct answer as fast as possible, or you will lose to the computer!
http://www.turtlediary.com/game/add-and-subtract-fractions.html

Frosty Fractions
Add together the two fractions given. If the answer is available on the board, place a snowflake token over it. The winner is the first player to get a straight line of three snowflakes, either horizontally, vertically, or diagonally (this game is for two players).
http://www.counton.org/games/map-fractions/frosty/

FRACTION MULTIPLICATION

Multiply Fractions by Whole Numbers
Use this simple online exercise for additional practice as needed.
http://www.mathgames.com/skill/4.67-multiply-fractions-by-whole-numbers

Multiply Fractions and Whole Numbers
Practice multiplying a whole number times a fraction in this online exercise.
https://www.khanacademy.org/math/pre-algebra/fractions-pre-alg/multiplying-fractions-pre-alg/e/multiplying_fractions_by_integers

Interactive Model for the Multiplication of Fractions
In this interactive activity, you will see how to use area models to multiply fractions.
https://www.learner.org/courses/learningmath/number/session9/part_a/try.html

Fraction Multiplication TeacherTool
Students multiply two fractions together and use an area model to represent the product. Scroll down to "Fifth Grade Multiplication and Division" and click on "Fraction Multiplication 2".
http://www.dreambox.com/teachertools

Soccer Math - Multiplying Fractions
Answer the multiple-choice fraction multiplication problems and play soccer in between the questions.
http://www.math-play.com/soccer-math-multiplying-fractions-game/multiplying-fractions-game.html

Snow Sprint Fractions
Practice fraction multiplication while participating in a snowmobile race!
http://www.mathplayground.com/ASB_SnowSprint.html

Multiply Fractions with Models
Use this simple online exercise for additional practice as necessary.
http://www.mathgames.com/skill/5.109-multiply-fractions-with-models

Product Fractions Card Activity (p. 36 of the PDF)
In this activity, players work in pairs to multiply fractions. This is not a "game", as such, but rather an opportunity for students to work collaboratively and manipulate the problems.
http://www.pepnonprofit.org/uploads/2/7/7/2/2772238/acing_math.pdf

Multiply Mixed Numbers Quiz
Practice multiplying mixed numbers. Express the answers as mixed numbers and in lowest terms.
https://www.thatquiz.org/tq-3/?-j304-la-p0

Fraction Multiplication Quiz
Practice multiplying like and unlike fractions in this 10-question interactive quiz.
http://www.thegreatmartinicompany.com/Math-Quick-Quiz/fraction-multiply-quiz.html

Multiplying Fractions Word Problems
Solve and interpret fraction multiplication word problems in this interactive exercise from Khan Academy.
https://www.khanacademy.org/math/arithmetic/fractions/multiplying-fractions-word-probl/e/multiplying-fractions-by-fractions-word-problems

Fraction Multiplication as Scaling
Interpret how multiplying by a fraction greater or less than 1 affects the product in this interactive online exercise.
https://www.khanacademy.org/math/pre-algebra/fractions-pre-alg/multiplying-fractions-pre-alg/e/fraction-multiplication-as-scaling

Multiplying Fractions Word Problems
Practice multiplying mixed numbers with these interactive word problems.
http://mrnussbaum.com/grade5standards/577-2/

Who Wants Pizza?
A tutorial that explains fraction multiplication using a pizza, followed by some interactive exercises.
http://math.rice.edu/~lanius/fractions/

Multiply Fractions Jeopardy
Jeopardy-style game. Choose a question by clicking on the tile that shows the points you will win.
http://www.quia.com/cb/95583.html

Multiplying Fractions
Multiply the fractions shown and reduce to the answer to the lowest terms.
http://www.mathplayground.com/fractions_mult.html

Fraction of a Number
Practice finding fractional parts of various numbers in this interactive online exercise.
https://www.mathplayground.com/fractions_fractionof.html

FRACTION DIVISION

Divide Fractions by Whole Numbers - Word Problems
These are simple word problems which can be used to reinforce the topic of division with fractions.
http://www.mathgames.com/skill/5.94-divide-fractions-by-whole-numbers

Dividing Unit Fractions by Whole Numbers
Use this simple interactive exercise to reinforce fraction division skills.
https://www.khanacademy.org/math/arithmetic/fractions/dividing-fractions-tutorial/e/dividing_fractions_0.5

Dividing Whole Numbers by Unit Fractions
Practice dividing a whole number by a unit fraction in this interactive exercise.
https://www.khanacademy.org/math/arithmetic/fractions/dividing-fractions-tutorial/e/dividing_fractions

Fractions as Divisions
In this educational video, Sal shows how a/b and a÷b are equivalent. That is, the fraction bar and the division symbol mean the same thing.
https://www.khanacademy.org/math/cc-fifth-grade-math/cc-5th-fractions-topic/tcc-5th-fractions-as-division/v/fractions-as-division

Partitive Division of Fractions Tool
This tool provides you with a connection between a story problem (context), paraphrase, model, and procedure where a fractional value is shared equally. Students complete phrases like: 5 4/5 shared equally among 12. They use a double number line or an area model to visualize the answer, and they connect this understanding to the numeric procedure.
https://www.conceptuamath.com/app/tool/divide-partitive

Fractions as Divisions Word Problems

Practice word problems that involve using the fraction bar as division in this interactive exercise.

https://www.khanacademy.org/math/cc-fifth-grade-math/cc-5th-fractions-topic/tcc-5th-fractions-as-division/e/understanding-fractions-as-division--word-problems

Dividing Fractions: Word Problems

Solve word problems by dividing fractions by fractions in this interactive exercise.

https://www.khanacademy.org/math/arithmetic/fractions/div-fractions-fractions/e/dividing-fractions-by-fractions-word-problems

Seven Cookies for Grampy

Make seven whole cookies for Grampy by rearranging the fractional parts of other whole cookies.

http://www.visualfractions.com/sevencookies/

Fraction Worksheets: Addition, Subtraction, Multiplication, and Division

Create custom-made worksheets for fraction addition, subtraction, multiplication, and division.

http://www.homeschoolmath.net/worksheets/fraction.php

Thinking Blocks: Ratio Word Problems

Model and solve word problems with ratios using this interactive bar model tool.

http://www.mathplayground.com/tb_ratios/index.html

GENERAL

Visual Fractions

A great site for studying all aspects of fractions, including: identifying, renaming, comparing, addition, subtraction, multiplication, division.

http://www.visualfractions.com/

Visual Fraction Calculator

This fraction calculator can perform addition, subtraction, multiplication, or division of two fractions. The values for the calculation can be simple or mixed fractions, or consist of only wholes. Input of improper fractions is also allowed.

http://www.dadsworksheets.com/fraction-calculator.html

Conceptua Math Fractions Tools

Free and interactive fraction tools. Each activity uses several models, such as circles, horizontal and vertical bars, number lines, etc. that allow students to develop a conceptual understanding of fractions.

http://www.conceptuamath.com/app/tool-library

Fraction Lessons at MathExpression.com

Tutorials, examples, and videos explaining all the basic fraction topics.

http://www.mathexpression.com/learning-fractions.html

Online Fraction Calculator

Add, subtract, multiply, or divide fractions and mixed numbers.

http://www.homeschoolmath.net/worksheets/fraction_calculator.php

Simplifying Fractions 1

Do you remember how to convert fractions into equivalent fractions?

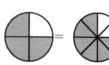

Each slice has been **split two ways**.

$$\frac{3}{4} = \frac{6}{8}$$

×2
×2

Each slice has been **split three ways**.

$$\frac{1}{3} = \frac{3}{9}$$

×3
×3

**We can also reverse the process.
Then it is called SIMPLIFYING or REDUCING a fraction:**

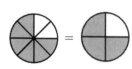

The slices have been **joined together in twos**.

$$\frac{6}{8} = \frac{3}{4}$$

÷2
÷2

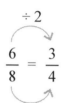

The slices have been **joined together in threes**.

$$\frac{3}{9} = \frac{1}{3}$$

÷3
÷3

Notice:

- Both the numerator and the denominator change into smaller numbers, but the **value of the fraction does not**. Think of it as getting the same amount of pie either way.

- After simplifying, the fraction is now written in a simpler form. We also say that the fraction is **reduced** or written in **lower terms**, because the new numerator and denominator are *smaller* numbers than the originals.

- Both the numerator and the denominator are *divided* by the same number.
  This number shows how many slices are joined together.

1. Write the simplification process, labeling the arrows. Follow the examples above.

a. The parts were joined together in _____.

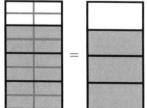

b. The parts were joined together in _____.

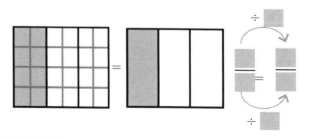

2. Write the simplifying process. You can write the arrows and the divisions to help you.

a. The slices were joined together in _____ .	**b.** The slices were joined together in _____ .	**c.** The slices were joined together in _____ .	**d.** The slices were joined together in _____ .
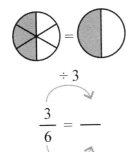	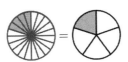	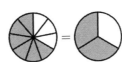	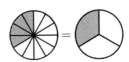
e. The slices were joined together in _____ .	**f.** The slices were joined together in _____ .	**g.** The slices were joined together in _____ .	**h.** The slices were joined together in _____ .
	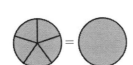	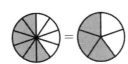	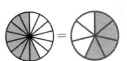

3. Draw a picture and reduce the fractions.

a. Join together each two parts.	**b.** Join together each four parts.	**c.** Join together each three parts.
	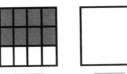 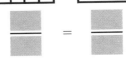	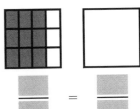
d. Join together each six parts.	**e.** Join together each seven parts.	**f.** Join together each four parts.
	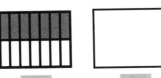 	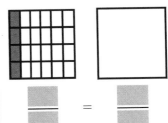

When we simplify a fraction, we need to *divide* the numerator and the denominator by some number, which means we need a number that "goes" into both the numerator and the denominator evenly—a number that is a **common factor** to both. Yet in other words, the numerator and the denominator have to be **divisible by some same number**.

Example 1. Reduce $\dfrac{28}{40}$.

Since both 28 and 40 are divisible by 4, we can divide the numerator and denominator by four. This means that each four slices are joined together.

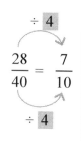

$$\dfrac{28}{40} = \dfrac{7}{10}$$

Example 2. Simplify $\dfrac{6}{17}$.

We cannot find any number that would go into 6 *and* 17 (except 1, of course). So 6/17 is already as simplified as it can be. It is already in its **lowest terms**.

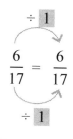

$$\dfrac{6}{17} = \dfrac{6}{17}$$

4. Reduce the fractions to lowest terms.

a. $\dfrac{6}{16} =$	**b.** $\dfrac{15}{25} =$	**c.** $\dfrac{3}{9} =$	**d.** $\dfrac{4}{8} =$	**e.** $\dfrac{16}{24} =$
f. $\dfrac{12}{20} =$	**g.** $\dfrac{24}{32} =$	**h.** $\dfrac{3}{15} =$	**i.** $\dfrac{15}{18} =$	**j.** $\dfrac{16}{20} =$

5. Simplify the fractional parts of these mixed numbers. The whole number does not change. Study the example.

a. $1\dfrac{4}{16} = 1\dfrac{1}{4}$	**b.** $5\dfrac{3}{27} =$	**c.** $7\dfrac{5}{20} =$	**d.** $3\dfrac{14}{49} =$

6. You cannot simplify some of these fractions because they are already in lowest terms. Cross out the fractions and mixed numbers that are already in lowest terms and simplify the rest.

a. $\dfrac{2}{3}$	**b.** $\dfrac{2}{6}$	**c.** $\dfrac{6}{13}$	**d.** $2\dfrac{7}{12}$
e. $\dfrac{11}{22}$	**f.** $5\dfrac{6}{12}$	**g.** $\dfrac{5}{11}$	**h.** $\dfrac{9}{20}$
i. $1\dfrac{4}{7}$	**j.** $3\dfrac{4}{28}$	**k.** $\dfrac{5}{29}$	**l.** $\dfrac{6}{33}$

7. Use a line to connect the fractions and mixed numbers that are equivalent.

$2\dfrac{6}{24}$ $\dfrac{28}{12}$ $1\dfrac{3}{4}$ $2\dfrac{4}{12}$

$\dfrac{14}{8}$ $2\dfrac{2}{8}$ $2\dfrac{5}{15}$ $\dfrac{21}{12}$

$2\dfrac{1}{3}$ $\dfrac{7}{4}$ $\dfrac{9}{4}$ $2\dfrac{1}{4}$

8. Tommy is on the track team. He spends 10 minutes warming up before practice and 10 minutes stretching after practice. All together, he spends a total of one hour for the warm-up, the practice, and the stretching.

What part of the total time is the warm-up time?

What part of the total time is the actual practice time?

9. Color slices of this 24-part circle according to how you spend your time during a typical day. Include sleeping, eating, bathing, school, housework, TV, *etc*. Write the name of each activity and what *fractional part* of your day it takes. Simplify the fractional part if you can.

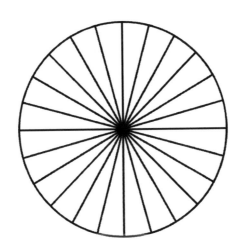

Simplifying Fractions 2

Example 1. Let's say we simplify $\frac{36}{48}$ into $\frac{6}{8}$.

Then, we could further simplify $\frac{6}{8}$ into $\frac{3}{4}$.

All of this could be done in *one* single step, if we divide 36 and 48 by 12 in the first place.

Since $\frac{3}{4}$ cannot be reduced any further, we say it is in its **lowest terms**.

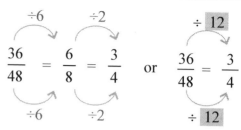

Whether you simplify a fraction in several steps or in one step, you get the same result.

When you cannot simplify a fraction any further, the fraction is in its *lowest terms*.

1. Simplify in two steps as indicated. Fill in the missing parts.

a.		b.	
You could simplify in one step if you divided by _____.	$\div 10 \quad \div 4$ $\frac{40}{120} = \underline{\quad} = \underline{\quad}$ $\div 10 \quad \div 4$	You could simplify in one step if you divided by _____.	$\div 5 \quad \div 3$ $\frac{75}{105} = \underline{\quad} = \underline{\quad}$ $\div 5 \quad \div 3$
c.		d.	
You could simplify in one step if you divided by _____.	$\div 3 \quad \div 2$ $\frac{30}{96} = \underline{\quad} = \underline{\quad}$ $\div 3 \quad \div 2$	You could simplify in one step if you divided by _____.	$\div 2 \quad \div 7$ $\frac{42}{98} = \underline{\quad} = \underline{\quad}$ $\div 2 \quad \div 7$

2. Simplify the following fractions.

a. $\frac{9}{27} =$	**b.** $\frac{9}{36} =$	**c.** $\frac{24}{32} =$	**d.** $\frac{42}{50} =$
e. $\frac{60}{200} =$	**f.** $\frac{24}{64} =$	**g.** $\frac{25}{70} =$	**h.** $\frac{27}{90} =$
i. $5\frac{20}{60} =$	**j.** $5\frac{66}{88} =$	**k.** $3\frac{16}{56} =$	**l.** $7\frac{36}{60} =$

3. Simplify these improper fractions using the same process you used for proper fractions. Then rewrite the simplified fraction as a mixed number. Follow the example.

a. $\dfrac{14}{8} = \dfrac{7}{4} = 1\dfrac{3}{4}$ ($\div 2$, $\div 2$)	**b.** $\dfrac{54}{36} =$	**c.** $\dfrac{32}{18} =$
d. $\dfrac{56}{49} =$	**e.** $\dfrac{64}{48} =$	**f.** $\dfrac{99}{54} =$

4. Three students did a problem that said, "Simplify to lowest terms."

$\dfrac{96}{120} = \dfrac{12}{\blacksquare} = \dfrac{\blacksquare}{\blacksquare}$	$\dfrac{96}{120} = \dfrac{8}{\blacksquare}$	$\dfrac{96}{120} = \dfrac{\blacksquare}{\blacksquare}$
a. Jerry divided the numerator and the denominator by 8, and then by 3.	**b.** Mark divided them by 12.	**c.** Nancy divided them by 24.

Who got it right? _____ Who didn't? _____

Why? _____

5. Some of the following fractions or fractional parts of mixed numbers *cannot* be simplified. Simplify the ones that can be simplified. Give your answer as a mixed number when possible.

a. $\dfrac{14}{29}$	**b.** $\dfrac{14}{22}$	**c.** $2\dfrac{8}{15}$
d. $\dfrac{8}{18}$	**e.** $\dfrac{22}{8}$	**f.** $\dfrac{27}{11}$
g. $4\dfrac{9}{23}$	**h.** $1\dfrac{21}{35}$	**i.** $\dfrac{44}{10}$

When solving problems with fractions, you should always give your answer:

- in lowest terms, and
- as a mixed number, if applicable.

$$\frac{4}{5} + \frac{7}{10} = \frac{8}{10} + \frac{7}{10} = \frac{15}{10} = \frac{3}{2} = 1\frac{1}{2}$$

change into   simplify   change into a
like fractions   mixed number

— OR —

It doesn't matter which you do first (convert into a mixed number or simplify) but usually it is easier to simplify first, then convert the fraction into a mixed number.

$$\frac{4}{5} + \frac{7}{10} = \frac{8}{10} + \frac{7}{10} = \frac{15}{10} = 1\frac{5}{10} = 1\frac{1}{2}$$

change into   change into a   simplify
like fractions   mixed number

6. Add or subtract. Give your answer as a mixed number and in lowest terms.

a. $\dfrac{5}{6} + 1\dfrac{2}{3}$	**b.** $\dfrac{7}{12} + \dfrac{3}{4}$
c. $\dfrac{3}{2} + \dfrac{6}{7}$	**d.** $\dfrac{9}{10} - \dfrac{1}{15}$
e. $\dfrac{7}{8} - \dfrac{5}{6}$	**f.** $\dfrac{15}{8} - \dfrac{3}{10}$
g. $3\dfrac{3}{4} - 1\dfrac{5}{6}$	**h.** $5\dfrac{5}{9} + 3\dfrac{7}{12}$

7. A computer screen is 1600 pixels wide. A horizontal line on the screen is 1200 pixels wide.

   a. What part of the width of the screen does the line take up?

   b. How wide should the line be if you want it to take up exactly 3/8 of the total width of the screen?

8. Simplify. Place the letter from each problem under the correct answer, and solve the riddle.

WHY ARE TEDDY BEARS NEVER HUNGRY?

$\frac{3}{4}$	$\frac{2}{5}$	$\frac{1}{2}$	$\frac{2}{7}$		$\frac{1}{4}$	$\frac{3}{5}$	$\frac{1}{2}$		$\frac{1}{4}$	$\frac{2}{3}$	$\frac{1}{6}$	$\frac{1}{4}$	$\frac{2}{7}$	$\frac{1}{3}$		$\frac{1}{3}$	$\frac{3}{4}$	$\frac{5}{6}$	$\frac{3}{10}$	$\frac{3}{10}$	$\frac{1}{2}$	$\frac{3}{7}$

Because ☐☐☐☐ ☐☐☐ ☐☐☐☐☐☐ ☐☐☐☐☐☐☐ .

E. $\dfrac{5}{10} =$	**H.** $\dfrac{4}{10} =$	**R.** $\dfrac{6}{10} =$	**E.** $\dfrac{18}{36} =$
L. $\dfrac{6}{9} =$	**S.** $\dfrac{12}{36} =$	**A.** $\dfrac{15}{60} =$	**S.** $\dfrac{24}{72} =$
W. $\dfrac{3}{18} =$	**E.** $\dfrac{30}{60} =$	**T.** $\dfrac{15}{20} =$	**Y.** $\dfrac{4}{14} =$
Y. $\dfrac{8}{28} =$	**F.** $\dfrac{15}{50} =$	**D.** $\dfrac{15}{35} =$	**A.** $\dfrac{20}{80} =$
F. $\dfrac{18}{60} =$	**T.** $\dfrac{12}{16} =$	**A.** $\dfrac{25}{100} =$	**U.** $\dfrac{25}{30} =$

Why do we need to give answers in lowest terms?

That is just considered the "tidy" way of writing fractional answers in mathematics. While it might not sound so important to you now, these principles are necessary in higher math, as well.

Consider the simplification of an *algebraic fraction* that you see on the right. We don't want to fail to simplify such an expression, seeing how simple it becomes in the end!

$$\frac{3x^2 - 3}{(1-x)(x^2+x)} = \frac{3(x^2-1)}{(1-x)x(x+1)} = \frac{3\cancel{(x-1)}(x+1)}{\underset{-1}{\cancel{(1-x)}}x\cancel{(x+1)}} = -\frac{3}{x}$$

The original expression.　　Factoring.　　Simplifying.　　The final answer.

Multiply Fractions by Whole Numbers

Example 1. $3 \times \frac{4}{5}$ is three copies of $\frac{4}{5}$.

How many fifths are there in total?

There are 12 fifths. So, $3 \times \frac{4}{5} = \frac{12}{5}$. Lastly we give the answer 12/5 as the mixed number 2 2/5.

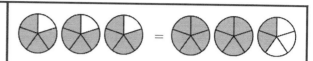

$$3 \times \frac{4}{5} = \frac{12}{5} = 2\frac{2}{5}$$

1. Repeatedly color parts to solve the multiplications. Give your answer as a <u>mixed number</u>.

a. $4 \times \frac{7}{9} =$

b. $3 \times \frac{5}{8} =$

c. $5 \times \frac{11}{12} =$

d. $6 \times \frac{7}{10} =$

2. Fill in.

a.

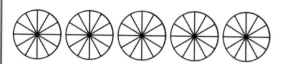

$2\frac{4}{5} = 2 \times \dfrac{}{}$

b.

$\dfrac{25}{9} = 5 \times \dfrac{}{}$

c.

$2\frac{2}{8} = 3 \times \dfrac{}{}$

You can draw illustrations to help you to visualize and solve the following problems.

3. Erica has tall drinking glasses that each hold 3/8 liters. How much water does she need to fill four of them?

4. Marlene wants to triple this recipe (make it three times). How much of each ingredient will she need?

<u>Brownies</u>

3/4 cup butter
1 1/2 cups brown sugar
4 eggs
1 1/4 cups cocoa powder
1/2 cup flour
2 tsp vanilla

To multiply a whole number and a fraction, find the total number of pieces.
This means you multiply the whole number and the top number (numerator) of the fraction.

Example 2. $8 \times \frac{3}{4}$ means 8×3 pieces, or 24 pieces. Each piece is a fourth. So, we get $\frac{24}{4}$.

Lastly, we write the answer as a mixed number. This time, $\frac{24}{4}$ happens to be the whole number 6.

Example 3. Multiplication can be done in either order. (In other words, multiplication is *commutative*.)

So, $\frac{3}{10} \times 5$ is the same as $5 \times \frac{3}{10}$. They both equal $\frac{5 \times 3}{10} = \frac{15}{10}$. This simplifies to $\frac{3}{2}$, which is $1\frac{1}{2}$.

5. Solve. Give your answer in lowest terms (simplified) and as a mixed number. Study the example.

a. $6 \times \frac{4}{9} = \frac{24}{9} = \frac{8}{3} = 2\frac{2}{3}$	**b.** $4 \times \frac{7}{10} =$
c. $2 \times \frac{11}{20} =$	**d.** $9 \times \frac{2}{15} =$
e. $\frac{15}{6} \times 2 =$	**f.** $6 \times \frac{7}{100} =$
g. $\frac{1}{12} \times 16 =$	**h.** $2 \times \frac{35}{100} =$
i. $\frac{9}{20} \times 10 =$	**j.** $\frac{7}{15} \times 7 =$

6. William asked 20 fifth graders how much time they spent on housework & chores the day before. He then rounded the answers to the nearest 1/8 hour. The line plot shows his results. Each x-mark corresponds to one fifth grader.

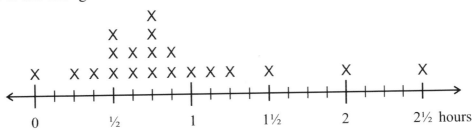

a. Exclude the three students who did the least housework and the three who did the most, and fill in:

Most students used between _____ and _____ hours for housework and chores.

b. The average for this data is 7/8 hours. Use this to calculate how many hours these 20 fifth graders used for housework in total.

111

A REMINDER		**Example 4.** $\frac{3}{10}$ of $120
A fraction _of_ a number means that **fraction TIMES the number.** In other words, the word "of" translates into multiplication.		$\downarrow$
		$\frac{3}{10} \times$ $120

You have previously learned how to find 3/10 of $120 using division:

- First, divide $120 by 10 to find 1/10 of it. It is $12.
- Then, multiply that by 3 to get 3/10 of $120. You get $36.

We also get the same answer with **fraction multiplication**: $\frac{3}{10} \times \$120 = \frac{3 \times \$120}{10} = \frac{\$360}{10} = \$36.$

Both methods are essentially the same: you divide by 10 and multiply by 3, just in two different orders.

7. Find the following quantities.

   a. 2/5 of 35 lb

   b. 4/9 of 180 km

8. Dad is building a shelf that is 4 meters long. He wants to use 2/5 of it for gardening supplies and the rest for tools. How long is the section of the shelf that is for gardening supplies? (_Hint: Use centimeters._)

9. **a.** Janet and Sandy earned $81 for doing yard work. They divided the money unequally so that Janet got 2/3 of it and Sandy got the rest. How much money did each girl get?

   b. What happens if the amount they earned is $80 instead?

10. Andy drew a 5 inch by 4 inch rectangle on paper. Then he drew a second rectangle that was 3/4 as long and wide as the first one.

    a. How long and how wide was Andy's second rectangle?

    b. Draw both rectangles (on separate paper).

Epilogue: There is something interesting about multiplying "a fraction times a whole number" and multiplying "a whole number times a fraction." Let's compare.

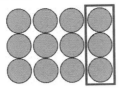 $\frac{1}{4} \times 12$ means **a fourth part of 12,** which is 3.	 $12 \times \frac{1}{4}$ means **12 copies of 1/4**, which makes 3 whole pies.

Notice: Both $\frac{1}{4} \times 12$ and $12 \times \frac{1}{4}$ equal 3. That makes sense, because multiplication can be done in any order. But they mean different things (a fourth part of 12, and 12 copies of 1/4).

11. Fill in the missing parts.

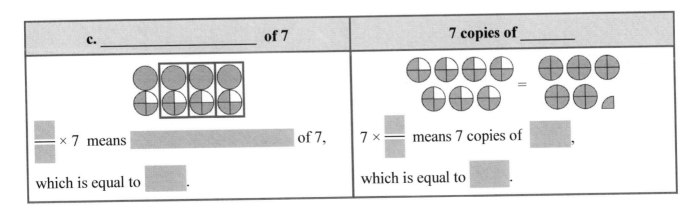

a. A two-fifth part of 10	**10 copies of 2/5**
$\frac{\boxed{}}{\boxed{}} \times 10$ means a two-fifth part of 10, which is equal to $\boxed{}$.	$10 \times \frac{\boxed{}}{\boxed{}}$ means 10 copies of $\boxed{}$, which is equal to $\boxed{}$.

b. A _____ part of 5	**5 copies of 1/3**
$\frac{1}{3} \times 5$ means a _____ part of 5, which is equal to $\boxed{}$.	$5 \times \frac{\boxed{}}{\boxed{}}$ means 5 copies of $\boxed{}$, which is equal to $\boxed{}$.

c. _____ of 7	**7 copies of _____**
$\frac{\boxed{}}{\boxed{}} \times 7$ means _____ of 7, which is equal to $\boxed{}$.	$7 \times \frac{\boxed{}}{\boxed{}}$ means 7 copies of $\boxed{}$, which is equal to $\boxed{}$.

113

Multiplying Fractions by Fractions

We have studied how to find a fractional part of a whole number using multiplication.

For example, $\frac{3}{5}$ of 80 is written as the multiplication $\frac{3}{5} \times 80 = \frac{240}{5} = 48$.

REMEMBER: The word "of" translates here into **multiplication**.

We can use the same idea to find a fractional part of a fraction.

Example 1. One-half of 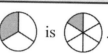 is .

As a multiplication, $\frac{1}{2} \times \frac{1}{3} = \frac{1}{6}$.

Example 2. One-fourth of is .

As a multiplication, $\frac{1}{4} \times \frac{1}{3} = \frac{1}{12}$.

1. Find a fractional part of the given fraction. You can think of a leftover pizza piece, which you must share equally with one, two, or three other people. Write a multiplication sentence.

a. Find $\frac{1}{2}$ of

$\frac{1}{2} \times \frac{1}{4} =$

b. Find $\frac{1}{2}$ of

 × =

c. Find $\frac{1}{2}$ of

 × =

d. Find $\frac{1}{3}$ of

e. Find $\frac{1}{3}$ of

f. Find $\frac{1}{3}$ of

g. Find $\frac{1}{4}$ of

h. Find $\frac{1}{4}$ of

i. Find $\frac{1}{4}$ of

Did you notice a shortcut?   If so, calculate $\frac{1}{5} \times \frac{1}{6} =$

114

<table>
<tr><td colspan="2" align="center">Shortcut: multiplying fractions of the type 1/n</td></tr>
<tr><td>To multiply fractions of the form 1/n where n is a whole number, simply multiply the denominators to get the new denominator.</td><td>$\rightarrow \quad \dfrac{1}{4} \times \dfrac{1}{5} = \dfrac{1}{20}$ or $\dfrac{1}{2} \times \dfrac{1}{6} = \dfrac{1}{12}$</td></tr>
</table>

2. Multiply.

a. $\dfrac{1}{9} \times \dfrac{1}{2}$	**b.** $\dfrac{1}{13} \times \dfrac{1}{3}$	**c.** $\dfrac{1}{5} \times \dfrac{1}{20}$

We have now studied how to find 1/2 or 1/3 or 1/5 of some fractions. What about finding some other kind of fractional part? Let's again compare this to finding fractional parts of whole numbers.

Review: To find $\dfrac{3}{4}$ of 16, or in other words $\dfrac{3}{4} \times 16$, you can first find $\dfrac{1}{4}$ of 16, which is 4.

Then just take that three times, which is 12. In other words, $\dfrac{3}{4} \times 16 = 12$.

We can use the same idea when finding a fractional part of another fraction.

Example 3. Find $\dfrac{2}{3}$ of $\dfrac{1}{4}$. First, we find $\dfrac{1}{3}$ of $\dfrac{1}{4}$, which is $\dfrac{1}{12}$.

Then, $\dfrac{2}{3}$ of $\dfrac{1}{4}$ is double that much, or $\dfrac{2}{12}$.

$\dfrac{2}{3}$ of ⬜ = ⬜

Example 4. Find $\dfrac{4}{5}$ of $\dfrac{1}{7}$.

First, we find $\dfrac{1}{5}$ of $\dfrac{1}{7}$, which is $\dfrac{1}{35}$. Then, $\dfrac{4}{5}$ of $\dfrac{1}{7}$ is four times that much, or $\dfrac{4}{35}$.

Multiplying a fraction by a fraction means taking that fractional part *of* the fraction. It is just like taking a certain part of the leftovers, when what is left over is a fraction.

3. The pictures show how much pizza is left, and you get a certain part of the leftovers. How much will you get? Color in a picture to show the answer.

a. $\dfrac{3}{4} \times$ ⬤ = ⬤	**b.** $\dfrac{2}{3} \times$ ⬤ = ⬤
c. $\dfrac{3}{4} \times$ ⬤ = ⬤	**d.** $\dfrac{2}{3} \times$ ⬤ = ⬤
e. $\dfrac{2}{5} \times$ ⬤ = ⬤	**f.** $\dfrac{4}{5} \times$ ⬤ = ⬤

4. Solve the multiplications by using two helping multiplications. Lastly, simplify if possible.

a. $\frac{2}{3} \times \frac{1}{8} =$	**b.** $\frac{3}{4} \times \frac{1}{10} =$
First find 1/3 of 1/8, then multiply the result by 2.	First find 1/4 of 1/10, then multiply the result by 3.
$\frac{1}{3} \times \frac{1}{8} = \frac{1}{24}$ and $\frac{1}{24} \times 2 = \underline{} = \underline{}$	$\frac{1}{4} \times \frac{1}{10} = \underline{}$ and $\underline{} \times 3 = \underline{}$
c. $\frac{3}{5} \times \frac{1}{6} =$	**d.** $\frac{5}{6} \times \frac{1}{9} =$
First find 1/5 of 1/6, then multiply the result by 3.	First find 1/6 of 1/9, then multiply the result by 5.
$\frac{1}{5} \times \frac{1}{6} = \underline{}$ and $\underline{} \times 3 = \underline{} = \underline{}$	$\frac{1}{6} \times \frac{1}{9} = \underline{}$ and $\underline{} \times 5 = \underline{}$
e. $\frac{2}{3} \times \frac{1}{7} =$	**f.** $\frac{3}{8} \times \frac{1}{4} =$

A shortcut for multiplying fractions

Multiply the numerators to get the numerator for the answer.
Multiply the denominators to get the denominator for the answer.

Study the examples on the right. Remember always to give your final answer <u>as a mixed number</u> and in lowest terms (simplified).	$\frac{3}{7} \times \frac{4}{9} = \frac{3 \times 4}{7 \times 9} = \frac{12}{63} = \frac{4}{21}$
	$\frac{4}{5} \times \frac{11}{8} = \frac{4 \times 11}{5 \times 8} = \frac{44}{40} = \frac{11}{10} = 1\frac{1}{10}$

5. Multiply. Give your answers in the lowest terms (simplified) and as mixed numbers, if possible.

a. $\frac{3}{9} \times \frac{2}{9}$	**b.** $\frac{11}{12} \times \frac{1}{6}$
c. $\frac{1}{3} \times \frac{3}{13}$	**d.** $9 \times \frac{2}{3}$
e. $\frac{2}{9} \times \frac{6}{7}$	**f.** $10 \times \frac{5}{7}$

COMPARE

The roundabout way	The shortcut
$\dfrac{5}{6} \times \dfrac{1}{2} = ?$ First find 1/6 of 1/2, then multiply the result by 5. $\dfrac{1}{6} \times \dfrac{1}{2} = \dfrac{1}{12}$ and $\dfrac{1}{12} \times 5 = \dfrac{5}{12}$	$\dfrac{5}{6} \times \dfrac{1}{2} = \dfrac{5 \times 1}{6 \times 2} = \dfrac{5}{12}$
$\dfrac{2}{8} \times \dfrac{3}{5} = ?$ Find 1/8 of 3/5, then multiply that result by 2. And to find 1/8 of 3/5, first find 1/8 of 1/5, and then multiply that by 3. $\dfrac{1}{8} \times \dfrac{1}{5} = \dfrac{1}{40}$. That multiplied by 3 is $\dfrac{1}{40} \times 3 = \dfrac{3}{40}$. Then, that multiplied by 2 is $\dfrac{3}{40} \times 2 = \dfrac{6}{40} = \dfrac{3}{20}$.	$\dfrac{2}{8} \times \dfrac{3}{5} = \dfrac{2 \times 3}{8 \times 5} = \dfrac{6}{40} = \dfrac{3}{20}$

In the "roundabout way," we do each multiplication separately.
In the shortcut, we can just do them all at once.

6. Multiply. Give your answers in the lowest terms (simplified) and as mixed numbers, if possible.

a.	$\dfrac{3}{4} \times \dfrac{7}{8} =$		b.	$\dfrac{7}{10} \times \dfrac{8}{5} =$
c.	$\dfrac{9}{20} \times \dfrac{4}{5} =$		d.	$\dfrac{2}{5} \times \dfrac{1}{3} =$
e.	$\dfrac{1}{4} \times \dfrac{2}{7} =$		f.	$\dfrac{9}{4} \times \dfrac{1}{3} =$
g.	$\dfrac{2}{3} \times \dfrac{11}{8} =$		h.	$\dfrac{2}{9} \times \dfrac{3}{10} =$

7. There was 1/4 of the pizza left. Marie ate 2/3 of that.

    a. What part of the *original* pizza did she eat?

    b. What part of the *original* pizza is left now?

8. Theresa has painted 5/8 of the room.

   a. What part is still left to paint?

   b. Now, Theresa has painted half of what was still left.
   Draw a bar model of the situation.
   What part of the room is still left to paint?

9. Ted has completed 2/3 of a job that his boss gave him.

   a. What part is still left to do?

   b. Now Ted has completed a third of what was still left to do.
   Draw a bar model of the situation.
   What (fractional) part of the original job is still left undone?

   What part is completed?

10. Sally wants to make 1/3 of the recipe on the right.
    How much does she need of each ingredient?

    Carob Brownies

    3 cups sweetened carob chips
    8 tablespoons extra virgin olive oil
    2 eggs
    1/2 cup honey
    1 teaspoon vanilla
    3/4 cup whole wheat flour
    3/4 teaspoon baking powder
    1 cup walnuts or other nuts

11. For an upcoming get-together, Alison needs to multiply the coffee
    recipe. Assume that half of the guests drink one serving, and the other
    half drink two servings. Find how much <u>coffee</u> she will need, if she has:

    Coffee (5 servings)

    3 1/2 cups water
    1/4 cup coffee

    a. 30 guests

    b. 50 guests

    c. 80 guests

Puzzle Corner   Find the missing factors.

a.	b.	c.	d.
$\times \dfrac{6}{7} = \dfrac{1}{7}$	$\times \dfrac{1}{4} = \dfrac{5}{16}$	$\times \dfrac{3}{8} = \dfrac{1}{16}$	$\times \dfrac{2}{5} = \dfrac{3}{10}$

Fraction Multiplication and Area

What is the area of this rectangle?

Notice, its side lengths are *fractional* (1/2 inch and 2/3 inch).

Let's extend its sides and draw a square inch around it.

Surely the area of our rectangle is less than a half square inch. But how much is the area exactly?

To solve this problem, let's draw a grid inside our square inch:

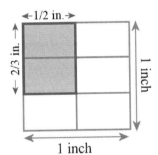

Now it is easy to see that the area of the colored rectangle is exactly 2/6 or 1/3 of the square inch.

(Why? Because the square inch is divided into 6 equal parts, and our rectangle covers two of them).

Notice that we get the same result (1/3 square inch) if we *multiply* the side lengths, using fraction multiplication:

$$\frac{2}{3} \text{ in} \times \frac{1}{2} \text{ in} = \frac{2}{6} \text{ in}^2 = \frac{1}{3} \text{ in}^2$$

1. Each picture shows some kind of square unit, and a colored rectangle. Figure out the side lengths and the area of the rectangle from the picture.

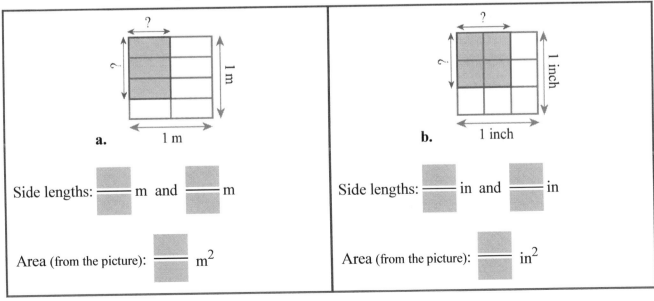

a.

Side lengths: ___ m and ___ m

Area (from the picture): ___ m²

b.

Side lengths: ___ in and ___ in

Area (from the picture): ___ in²

2. Again, figure out the side lengths of the colored rectangle from the picture. Then multiply the side lengths to find its area. <u>Check that the area you get by multiplying is the same as what you can see from the picture.</u>

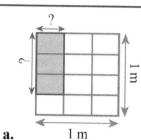

a. 1 m

Side lengths: 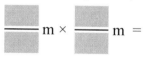 m and m

Area (by multiplication):

m × m =

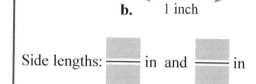

b. 1 inch

Side lengths:  in and in

Area (by multiplication):

in × in =

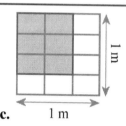

c. 1 m

Side lengths: 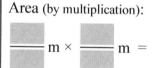 m and m

Area (by multiplication):

m × m =

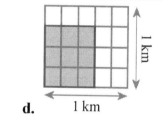

d. 1 km

Side lengths: 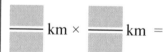 km and km

Area (by multiplication):

km × km =

3. Shade a rectangle inside the square so that its area can be found by the fraction multiplication.

a.  $\frac{1}{4}$ m $\times$ $\frac{1}{2}$ m $=$ $\frac{1}{8}$ m^2

b.  $\frac{1}{2}$ in $\times$ $\frac{4}{6}$ in $=$ $\frac{4}{12}$ in^2

c.  $\frac{3}{4}$ ft $\times$ $\frac{2}{7}$ ft $=$

d. 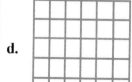 $\frac{3}{5}$ km $\times$ $\frac{5}{6}$ km $=$

The area of this rectangle *can* be found by multiplication:

$\frac{3}{4}$ m $\times$ $\frac{1}{3}$ m $=$ $\frac{1}{4}$ m²; however, we want to <u>verify</u> this using a visual method.

For that reason, let's sketch a unit square around the rectangle and tile it.

We need to extend the sides of the rectangle to draw the square. The 1/3-meter side simply needs to be three times as long to make it 1 meter.

Then, divide the side that is 3/4 meters long into three equal parts— each part is 1/4 m long. Then extend that side by another 1/4 meter.

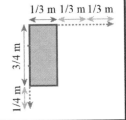

Lastly, draw the entire square. Draw gridlines to show the tiles within the square meter: one side is divided into 3 equal parts, and the other into 4 equal parts. We get 12 tiles.

Now it is easy to see that the area of the colored rectangle is 3 tiles out of 12, or 3/12 of a square meter. That simplifies to 1/4 of a square meter.

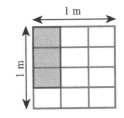

---

4. Extend the sides of the rectangle so you get a square meter (unit square). Draw gridlines into the square as in the example above. Write a multiplication for the area of the colored rectangle. <u>Verify</u> <u>that the area you get by multiplying is the same as what you can see in the picture.</u>

a. 1/3 m

1/3 m

Area: ▭/▭ m × ▭/▭ m =

b. 1/5 m

1/3 m

c. 1/5 m

1/2 m

d. 1/4 m

1/4 m

5. Extend the sides of the rectangle so you get a square meter (unit square). Draw gridlines into the square as in the example above. Write a multiplication for the area of the colored rectangle. Verify that the area you get by multiplying is the same as what you can see in the picture.

a. 3/4 m

1/2 m

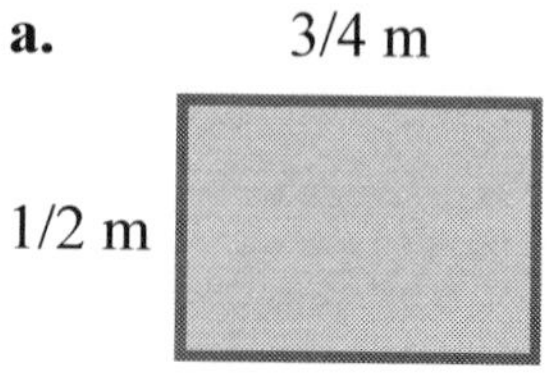

b. 2/5 m

3/4 m

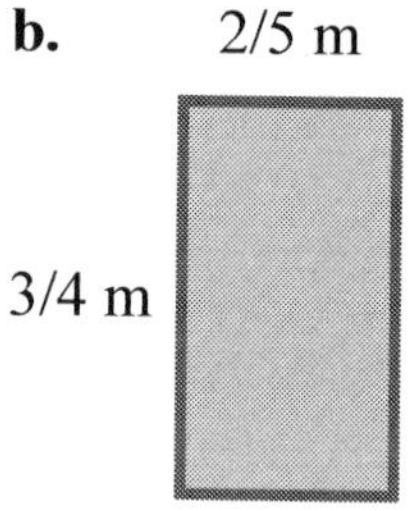

c. 2/3 m

2/3 m

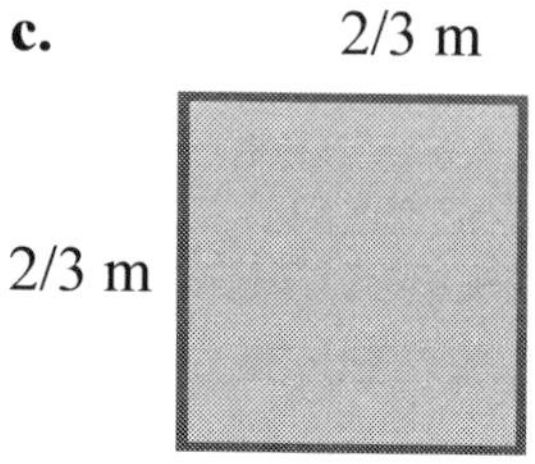

d. 3/5 m

1/2 m

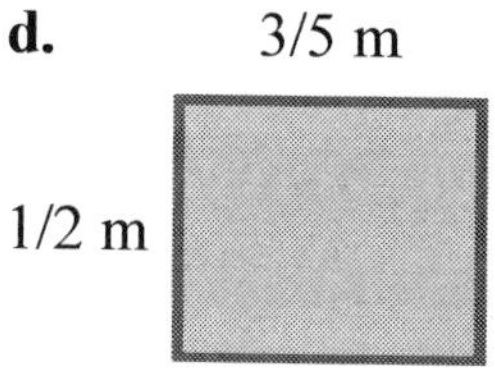

e. 3/4 m

3/4 m

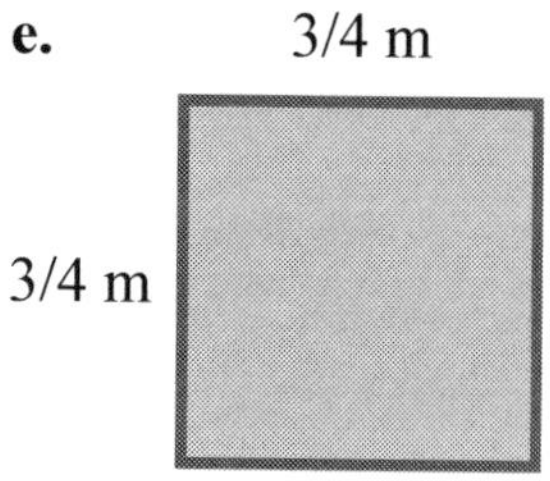

f. 5/6 m

1/2 m

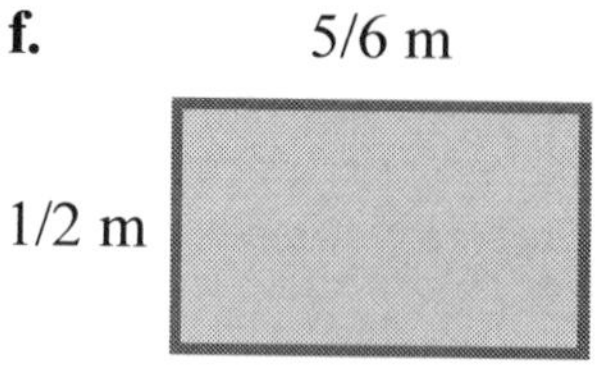

6. In the pictures below, the outer square is <u>one square unit</u>. Write a multiplication for the area of the colored rectangle. This time, we are not using meters or inches, just "units" and "square units," and you do not have to include those in the multiplication (simply write the fractions without any units).

a. ____ / ____ × ____ / ____ =

b. ____ / ____ × ____ / ____ =

c. ____ / ____ × ____ / ____ =

d. ____ / ____ × ____ / ____ =

7. **a.** Draw a 1 in by 1 in square. What is its area?

b. Draw a rectangle with 3/4 in and 5/8 in sides inside the square you drew so that the two sides of the rectangle touch the sides of the square. See the illustration below (not to scale).

c. Find the area of your rectangle.

8. **a.** Draw a square centimeter.

   b. Draw the rectangle with 3/10 cm and 7/10 cm sides
   inside the square centimeter so that the two sides
   of the rectangle touch the sides of the square.

   c. Calculate the area of the rectangle in square centimeters
   using both fractions and decimals (calculate it two times).
   Using fractions:

   Using decimals:

9. **a.** Find the area of a rectangular suburb that is 3 km by 1/2 km.

   b. A village lies inside a 5/8 mile by 3/4 mile rectangle. Find its area.

10. **a.** A stamp measures 7/8 in by 3/4 in. Amanda
    puts 6 of them onto an envelope, side by side.
    Find the total area these stamps cover.

    b. The envelope is 8 in by 5 in.
    About what part of the envelope do the six stamps cover?

Puzzle Corner   Which has a larger area, a square with 7/8-mile sides,
or a rectangle that is 1/4 mile by 3 miles?

Simplifying Before Multiplying

A new notation

We will start using a new way to indicate simplifying fractions. When a numerator or a denominator gets simplified, we will cross it out with a slash and write the *new* numerator or denominator next to it (either above it or below it).

The number you divide by (the 4) does *not* get indicated in any way! You only think about it in your mind: "I divide 12 by 4, and get 3. I divide 20 by 4, and get 5."

You may not see any advantage over the "old" method yet, but this shortcut will come in handy soon.

$$\frac{\cancel{12}^{\,3}}{\cancel{20}_{\,5}} = \frac{3}{5}$$

$$\frac{\cancel{35}^{\,7}}{\cancel{55}_{\,11}} = \frac{7}{11}$$

1. Simplify the fractions. Write the simplified numerator and denominator above and below the old ones.

a. $\dfrac{14}{16}$	b. $\dfrac{33}{27}$	c. $\dfrac{12}{26}$	d. $\dfrac{9}{33}$

Before multiplying, we can write another equivalent, simpler fraction in the place of a fraction.

In the first example on the right, 3/6 is simplified to 1/2 before multiplying. We write a tiny "1" above the "3" and a tiny "2" below the "6". In the other example, 4/10 is simplified to 2/5 before multiplying.

Why does this work? Obviously, we can write 1/2 instead of 3/6, or 2/5 instead of 4/10, since they are *equivalent*.

Example 1:
$$\frac{\cancel{3}^{\,1}}{\cancel{6}_{\,2}} \times \frac{5}{8} = \frac{5}{16}$$

Example 2:
$$\frac{3}{7} \times \frac{\cancel{4}^{\,2}}{\cancel{10}_{\,5}} = \frac{6}{35}$$

2. Simplify one or both fractions before multiplying. Use equivalent fractions. Look at the example.

a. $\dfrac{\cancel{6}^{\,3}}{\cancel{10}_{\,5}} \times \dfrac{\cancel{2}^{\,1}}{\cancel{14}_{\,7}} = \dfrac{3 \times 1}{5 \times 7} = \dfrac{3}{35}$	b. $\dfrac{2}{4} \times \dfrac{3}{15} =$
c. $\dfrac{8}{12} \times \dfrac{1}{2} =$	d. $\dfrac{8}{32} \times \dfrac{14}{21} =$
e. $\dfrac{6}{15} \times \dfrac{6}{9} =$	f. $\dfrac{27}{45} \times \dfrac{21}{49} =$

You can also simplify "criss-cross." Look at the example on the right: →

We simplify 3 and 6, writing 1 and 2 in their place. Think of it as the fraction 3/6 being simplified into 1/2, but the 3 and 6 are across from each other.

$$\dfrac{7}{\overset{}{\underset{2}{6}}} \times \dfrac{\overset{1}{3}}{9} = \dfrac{7}{18}$$

Why are we allowed to simplify in such a manner?

Compare the above problem to this one: $\dfrac{7}{9} \times \dfrac{3}{6}$. (It is almost the same, isn't it?) Surely you can see that in this problem, we *could* simplify 3/6 to 1/2 before multiplying.

And, these two multiplication problems are essentially the *same* problem, because they both lead to the same expression and the same answer: the first one becomes $\dfrac{7 \times 3}{6 \times 9} = \dfrac{21}{54}$, and the second one becomes $\dfrac{7 \times 3}{9 \times 6} = \dfrac{21}{54}$ (without simplifying). Therefore, since you can simplify 3/6 into 1/2 in the one problem, you can do the same in the other also.

3. Simplify "criss-cross" before you multiply.

a. $\dfrac{8}{9} \times \dfrac{6}{11}$	**b.** $\dfrac{3}{10} \times \dfrac{2}{5}$	**c.** $\dfrac{4}{7} \times \dfrac{1}{12}$
d. $\dfrac{7}{4} \times \dfrac{3}{21}$	**e.** $\dfrac{3}{16} \times \dfrac{8}{5}$	**f.** $\dfrac{3}{8} \times \dfrac{12}{11}$

You can even simplify criss-cross several times before multiplying.	$\dfrac{\overset{1}{3}}{15} \times \dfrac{5}{\underset{2}{6}}$ First, simplify 3 and 6 into 1 and 2.	$\dfrac{\overset{1}{3}}{\underset{3}{15}} \times \dfrac{\overset{1}{5}}{\underset{2}{6}} = \dfrac{1}{6}$ Then simplify 5 and 15 into 1 and 3.

4. Simplify before you multiply.

a. $\dfrac{7}{8} \times \dfrac{2}{7}$	**b.** $\dfrac{3}{5} \times \dfrac{5}{6}$	**c.** $\dfrac{5}{12} \times \dfrac{4}{10}$
d. $\dfrac{9}{15} \times \dfrac{3}{18}$	**e.** $\dfrac{8}{11} \times \dfrac{3}{4}$	**f.** $\dfrac{12}{100} \times \dfrac{4}{15}$

Example 3. Simplify $\frac{27}{45} \times 45$. You can think of this problem in two manners:

(1) Think of the fraction line as division. The problem is therefore the same as $27 \div 45 \times 45$. Whenever you multiply and divide by the same number, they cancel each other out. So, you can cross out both 45s in the original problem, and the answer is simply 27.

(2) First change the whole number 45 into the fraction 45/1. The problem is now $\frac{27}{45} \times \frac{45}{1}$.

Now you can simplify criss-cross, and multiply: $\frac{27}{\cancel{45}_1} \times \frac{\cancel{45}^1}{1} = 27$.

5. Simplify and multiply.

a. $\frac{82}{77} \times 77 =$	**b.** $13 \times \frac{49}{13} =$	**c.** $\frac{14 \times 16}{14} =$
d. $\frac{5}{6} \times 24 =$	**e.** $54 \times \frac{2}{9} =$	**f.** $\frac{16 \times 5}{8} =$

6. A toy block is 3/8 in tall. How tall is a stack of 8 of them?

   A stack of 20 of them?

7. Sandra buys 3/4 kg of meat every week. How much meat does she buy in a year?

8. The morning after Sam's birthday, 12/20 of his birthday cake is left. He eats 2/3 of what is left. When you multiply those two fractions, what does your answer mean or tell you?

To multiply three or more fractions, the same principles apply. You multiply all the numerators and all the denominators to get the numerator and the denominator for the answer.

Example 4. We can do a lot of simplifying before multiplying with this problem: $\dfrac{14}{25} \times \dfrac{10}{9} \times \dfrac{5}{6}$

$\dfrac{14}{\cancel{25}^{}} \times \dfrac{\cancel{10}^{2}}{9} \times \dfrac{5}{6}$ $_5$	$\dfrac{\cancel{14}^{7}}{\cancel{25}_5} \times \dfrac{\cancel{10}^{2}}{9} \times \dfrac{5}{\cancel{6}_3}$	$\dfrac{\cancel{14}^{7}}{\cancel{25}_{\cancel{5}_1}} \times \dfrac{\cancel{10}^{2}}{9} \times \dfrac{\cancel{5}^{1}}{\cancel{6}_3} \;=\; \dfrac{14}{27}$
1. Simplify 10 and 25 into 2 and 5 (dividing by 5).	2. Simplify 14 and 6 into 7 and 3.	3. Lastly, simplify 5 and 5, leaving 1 and 1.

9. Multiply three fractions. Simplify before multiplying.

a. $\dfrac{4}{5} \times \dfrac{3}{4} \times \dfrac{2}{3} =$	**b.** $\dfrac{11}{8} \times \dfrac{6}{8} \times \dfrac{2}{3} =$
c. $\dfrac{9}{10} \times \dfrac{5}{2} \times \dfrac{2}{7} =$	**d.** $\dfrac{3}{5} \times \dfrac{6}{12} \times \dfrac{5}{3} =$
e. $\dfrac{4}{5} \times \dfrac{9}{8} \times \dfrac{10}{24} =$	**f.** $\dfrac{7}{12} \times \dfrac{3}{5} \times \dfrac{6}{7} =$

10. **a.** Draw a bar model for this situation. Matthew pays 1/5 of his salary in taxes. Of what is left, he uses 1/4 to purchase groceries.

b. Suppose Matthew's salary is $2,450. Calculate how much he uses for groceries.

| **Epilogue: What happens if you *don't* simplify before multiplying?** Compare the two problems on the right →

 Jack did all of the simplifying before multiplying. Tina simplified after multiplying. Both of them got the right answer. Simplifying before multiplying does NOT change the final answer—it just makes it <u>easier to multiply</u> because the numbers are *smaller*! | $\dfrac{7}{35} \times \dfrac{6}{8} = \dfrac{42}{280} = \dfrac{21}{140} = \dfrac{3}{20}$

 Tina multiplies first to get 42/280. Lastly, she simplifies her *answer* in two steps, first to 21/140, and then to 3/20. | $\dfrac{\cancel{7}^{1}}{\cancel{35}_5} \times \dfrac{\cancel{6}^{3}}{\cancel{8}_4} = \dfrac{3}{20}$

 Jack simplifies before multiplying. |

Multiplying Mixed Numbers

Multiplying mixed numbers is not difficult at all.

- First, change the mixed numbers to fractions.
- Then multiply the fractions.
- Give your answer as a mixed number and in lowest terms. Also, check that your answer makes sense (by estimation).

The most difficult part of all this is to **remember *not* to multiply the mixed numbers until you have first changed them into fractions!**

$$1\frac{2}{3} \times 2\frac{5}{6}$$

$$\downarrow \qquad \downarrow$$

$$\frac{5}{3} \times \frac{17}{6} = \frac{85}{18} = 4\frac{13}{18}$$

Estimation: $1\ 2/3 \times 3 = 5$. The answer is fairly close to 5, so it is reasonable.

1. First change the mixed numbers to fractions. Then multiply. Give your answer as a mixed number, with the fractional part in lowest terms. You can use estimation to check if your answer is reasonable.

a. $2\frac{1}{4} \times 1\frac{1}{2}$

$\downarrow \qquad \downarrow$

b. $10\frac{1}{3} \times 2\frac{1}{2}$

c. $5\frac{1}{5} \times \frac{1}{6}$

d. $4\frac{1}{2} \times 3\frac{1}{5}$

It helps to simplify before you multiply, because then the numerators and the denominators you multiply are smaller. Also, this makes it easier to convert the final answer into a mixed number.

Estimation: $4 \times (3\ 1/2) = 14$. The answer 14 ¼ is close to that, so it makes sense.

$$4\frac{2}{9} \times 3\frac{3}{8}$$

$$\downarrow \qquad \downarrow$$

$$\frac{\overset{19}{\cancel{38}}}{\underset{1}{\cancel{9}}} \times \frac{\overset{3}{\cancel{27}}}{\underset{4}{\cancel{8}}} = \frac{57}{4} = 14\frac{1}{4}$$

e. $3\frac{5}{6} \times 3\frac{1}{3}$

f. $3\frac{1}{3} \times 5\frac{1}{10}$

129

2. Practice some more. You can use estimation to check if your answers are reasonable.

a. $2\frac{3}{5} \times 1\frac{1}{6}$	**b.** $3\frac{3}{10} \times 2\frac{1}{3}$
c. $1\frac{1}{8} \times 2\frac{4}{9}$	**d.** $3\frac{3}{4} \times 3\frac{1}{3}$

3. **a.** A carpet is 5 ½ feet wide and 7 ½ feet long. How many square feet does it cover?

   b. A room is 12 ft by 20 ft. *About* what part of the floor area does the carpet cover? Use estimation (rounded numbers).

4. Alice is going to make this recipe 1 ½ times. Calculate the correct amount of each ingredient for her. Write the new amounts on the lines in front of the numbers in the recipe.

   Cheeseball

   _____ 2 packages cream cheese

   _____ 2 ½ cups shredded Cheddar cheese

   _____ 1 ½ cups chopped pecans

   _____ 1 teaspoon grated onion

5. Is it possible to multiply mixed numbers in parts like Nathan does? Check his calculations this way: change the mixed number into a fraction first, then multiply.

a. Nathan's way (multiply in parts): $5 \times 4\frac{1}{7} = 5 \times 4 + 5 \times \frac{1}{7}$ $= 20 + \frac{5}{7} = 20\frac{5}{7}$	**a.** Change the mixed number into a fraction first: $5 \times 4\frac{1}{7}$
b. Nathan's way (multiply in parts): $8 \times 2\frac{5}{6} = 8 \times 2 + 8 \times \frac{5}{6}$ $= 16 + \frac{\overset{4}{\cancel{8}}}{1} \times \frac{5}{\underset{3}{\cancel{6}}} = 16\frac{20}{3} = 22\frac{2}{3}$	**b.** Change the mixed number into a fraction first: $8 \times 2\frac{5}{6}$

6. Correct Dylan's calculation for the problem 1 ½ × 1 ½ .

Dylan thought: "I will just multiply 1 × 1 = 1, and ½ × ½ = ¼, and add them. I get 1 ¼ ."

Now, Dylan's method is <u>wrong</u>! Just think: multiplication by 1 ½ should make the number you multiply (1 ½) bigger, not smaller! Find the correct answer.

Where did Dylan go wrong? To see that, study the picture on the right. The *outer* square is 2 units by 2 units. Thinking in meters, for example, the outer square is 2 meters by 2 meters.

The colored square is 1 ½ m by 1 ½ m. The area of that square is found by multiplying 1 ½ × 1 ½ .

The colored square is divided into four parts, marked with 1, 2, 3, and 4.

Can you tell their areas by looking at the picture?

Part 1 is one square meter, part 2 is ½ square meter, part 3 is ½ square meter, and part 4 is ¼ square meter. Adding those, we get a total area of 2 ¼ square meters. So, 1 ½ × 1 ½ equals 2 ¼, not 1 ¼.

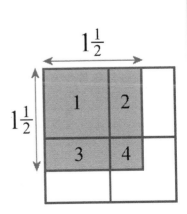

The *correct* way to multiply mixed numbers in parts

For two mixed numbers, there are *four* partial multiplications to do. Let's solve 1 ½ × 2 ½ for an example.

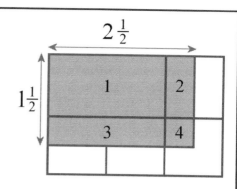

1. Multiply the two whole-number parts: 1 × 2 = 2.
2. Multiply the first whole number times the second fractional part: 1 × ½ = ½.
3. Multiply the first fractional part times the second whole number: ½ × 2 = 1.
4. Multiply the two fractional parts: ½ × ½ = ¼.

Then add the four partial results to get 3 ¾. These four parts are marked in the picture, and the partial products correspond to the *areas* of the four parts.

Even though all of this is a lot of work, it may *still* be less work than changing mixed numbers into fractions, multiplying them, and then changing the answer back into a mixed number. It is your choice which method you prefer to use.

7. Multiply using the regular method, or in parts.

a. $2 \times 7\frac{1}{3}$	**b.** $2\frac{1}{9} \times 5\frac{1}{3}$
c. $7 \times 2\frac{4}{7}$	**d.** $4 \times 3\frac{2}{9}$
e. $2\frac{5}{8} \times 3\frac{2}{3}$	**f.** $2\frac{1}{2} \times 2\frac{1}{5}$

8. In the US "letter" size paper measures 8 1/2 × 11 inches.

    a. What is the area of this kind of paper in square inches?

    b. If you use ½-inch margins on all four sides,
       what is the real writing area in square inches?
       Hint: Make a sketch of the situation.

9. Samantha is making a quilt with squares like the one on the right.
The big square measures 6 in × 6 in and the square inside it
measures 3 ¼ in × 3 ¼ in.

    a. Calculate the area of the inside square.

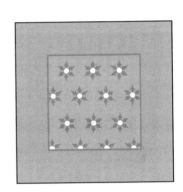

    b. Does the inside square take up more or less than
       1/3 of the area of the outer square?

Multiplication as Scaling/Resizing

You have learned that:

- The multiplication (2/3) × 18 km can be thought of as finding two-thirds of 18 km.
- The multiplication (3/4) × (1/2) can be thought of as finding three-fourths of 1/2 (say, of a pie).

In general, when you multiply a fraction times another number, think of it as finding a fractional part of the other number. In this sense, **the symbol "×" translates to "of."**

We can use the same idea when we multiply a mixed number times a number.

Example 1. The multiplication (1 2/3) × 18 km is the same as 1 times 18 km *plus* two-thirds of 18 km. In essence, we multiply in parts.

Since two-thirds of 18 km is 12 km, then (1 2/3) × 18 km is 18 km + 12 km = 30 km.

1. Write as a multiplication using a fraction, and solve. Remember, "of" translates into "×." Use the top problem in each box to help you solve the bottom one.

a. one-fourth of 24	**b.** one-tenth of 110 kg	**c.** one-fifth of one-half
☐/☐ × ___24___ = _____	☐/☐ × _____ = _____	☐/☐ × ☐/☐ = _____
three-fourths of 24	nine-tenths of 110 kg	three-fifths of one-half
☐/☐ × _____ = _____	☐/☐ × _____ = _____	☐/☐ × ☐/☐ = _____

2. Write as a multiplication using a fraction or a mixed number, and solve. Use the top problem in each box to help you solve the bottom ones.

a. one-fourth of 28	**b.** one-sixth of 12 mi	**c.** one-fifth of $400
☐/☐ × ___28___ = _____ 	☐/☐ × _____ = _____ 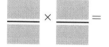	☐/☐ × _____ = _____ 
one and one-fourth times 28	two and one-sixth times 12 mi	two-fifths of $400
☐☐/☐ × _____ = _____ 	☐☐/☐ × _____ = _____ 	☐☐/☐ × _____ = _____ 
two and one-fourth times 28	six and one-sixth times 12 mi	four and two-fifths times $400
☐☐/☐ × _____ = _____ 	☐☐/☐ × _____ = _____ 	☐☐/☐ × _____ = _____ 

You also know that **scaling means expanding or shrinking something by some factor**.

Example 2. This stick ——— is 40 pixels long. Let's scale it to be 6/10 as long as it was (it will shrink!):

——— → ——

From this, we can write a multiplication: $(6/10) \times$ ——— = —— , or $(6/10) \times 40$ px = 24 px. We can even use decimals, and write 0.6×40 px = 24 px.

The number we multiply by is called the **scaling factor**.

If the scaling factor is more than 1, such as 2 2/3, the resulting stick is *longer* than the original one. If the scaling factor is less than 1, such as 1/6 or 7/8, the resulting stick is *shorter*.

3. The stick is being scaled—either expanded or shrunk. How long will it be in pixels? Compare the problems in each box.

a.	b.	c.
½ × ——— = ——	¼ × ——— = —	¾ × ——— = ——
½ × 50 px = _____ px	¼ × 40 px = _____ px	¾ × 48 px = _____ px
1 ½ × ——— = ——	2 ¼ × ——— = ———	1 ¾ × ——— = ———
1 ½ × 50 px = _____ px	2 ¼ × 40 px = _____ px	1 ¾ × 48 px = _____ px

d. $\frac{5}{8}$ × ——— = ——	e. $\frac{3}{5}$ × ——— = ——
$\frac{5}{8}$ × 40 px = _____ px	$\frac{3}{5}$ × 50 px = _____ px
2 $\frac{5}{8}$ × ——— = ———————	3 $\frac{3}{5}$ × ——— = ———————
2 $\frac{5}{8}$ × 40 px = _____ px	3 $\frac{3}{5}$ × 50 px = _____ px

4. Will the resulting stick be longer or shorter than the original—or equally long? You do not have to calculate anything. Compare.

a. $\frac{2}{3}$ × ——— is longer/shorter than ———.	b. $1\frac{2}{3}$ × ——— is longer/shorter than ———.
c. $\frac{9}{8}$ × ——— is longer/shorter than ———.	d. $\frac{3}{7}$ × ——— is longer/shorter than ———.
e. $3\frac{2}{100}$ × ——— is longer/shorter than ———.	f. $\frac{99}{100}$ × ——— is longer/shorter than ———.

135

Nuts cost $9 per pound. How do you find the price of 2/3 of a pound of nuts?

You multiply the price by 2/3.
Or, you can divide by 3 (to find 1/3 of the price), and then multiply that by 2. Either way will work!

5. Find the total cost.

    a. Fuel costs $6 per gallon, and you buy 3/4 of a gallon.

    b. Nuts cost 8 ½ dollars per pound. You buy 1 ½ pounds.

    c. Chain costs $11.30 per meter, and you buy 2.6 meters.
        (Hint: Calculate with decimals in this case, naturally!)

    d. Rent is $350 per month (30 days). You stay for 12 days.
        (Hint: If you use fraction multiplication, simplify your fraction first.)

6. Chloe had a photograph on the computer that was 2,400 pixels wide and 1,600 pixels tall. She resized it so that its width and height became 2/3 of that. Calculate the new image dimensions, rounded to the nearest whole number.

7. Mason drew an image on the computer. It was 360 pixels by 600 pixels. Then he expanded it to be 2 ¼ times as wide and tall. What are the new dimensions of his image?

A neat connection

You have learned to use multiplication with equivalent fractions: $\dfrac{3}{4} = \dfrac{15}{20}$ $\overset{\times\,5}{\curvearrowright}$ $\underset{\times\,5}{\curvearrowleft}$

We can write the same process this way also: $\dfrac{3}{4} = \dfrac{5 \times 3}{5 \times 4} = \dfrac{15}{20}$

Now notice: $\dfrac{5 \times 3}{5 \times 4}$ is the same as $\dfrac{5}{5} \times \dfrac{3}{4}$, isn't it? And $\dfrac{5}{5}$ is equal to 1.

Therefore, $\dfrac{5 \times 3}{5 \times 4}$ is actually the same as multiplying $\dfrac{3}{4}$ by 1.

So, we can make equivalent fractions by multiplying a given fraction by 1—we just write 1 in the form of a fraction (such as 3/3).

8. Make equivalent fractions by multiplying the given fraction by different forms of the number 1.

a. Multiply the given fraction by $\dfrac{4}{4}$. $\dfrac{\boxed{4}}{\boxed{4}} \times \dfrac{2}{3} =$	**b.** Multiply the given fraction by $\dfrac{3}{3}$. $\dfrac{}{} \times \dfrac{5}{9} =$
c. Multiply the given fraction by $\dfrac{10}{10}$. $\dfrac{}{} \times \dfrac{2}{7} =$	**d.** Multiply the given fraction by $\dfrac{7}{7}$. $\dfrac{}{} \times \dfrac{11}{12} =$

9. Is the result of multiplication more, less, or equal to the original number? You do not have to calculate anything. Compare and write $<$, $>$, or $=$ in the box.

a. $\dfrac{9}{10} \times 16$ ☐ 16	**b.** $5\dfrac{7}{9} \times 31$ ☐ 31	**c.** $\dfrac{6}{6} \times 5$ ☐ 5
d. $\dfrac{20}{20} \times 88$ ☐ 88	**e.** $\dfrac{4}{5} \times \dfrac{2}{3}$ ☐ $\dfrac{2}{3}$	**f.** $\dfrac{11}{4} \times 164$ ☐ 164
g. $\dfrac{7}{5} \times \dfrac{4}{4}$ ☐ $\dfrac{7}{5}$	**h.** 2.61×7 ☐ 7	**i.** 0.918×431 ☐ 431

Puzzle Corner

Find the expressions that are equivalent to 5/8.

a. $\dfrac{3 \times 5}{6 \times 4}$	**b.** $\dfrac{9 \times 5}{8 \times 9}$	**c.** $\dfrac{5}{6} \times \dfrac{6}{5}$	**d.** $\dfrac{5}{7} \times \dfrac{7}{8}$	**e.** $\dfrac{2}{8} \times \dfrac{5}{3} \times \dfrac{3}{2}$	**f.** $\dfrac{4 \times 9 \times 5}{5 \times 9}$	**g.** $\dfrac{2 \times 5 \times 8}{8 \times 5 \times 8}$

Fractions Are Divisions

Each fraction is also a division problem.

Example 1. $\frac{2}{5}$ is also the division problem $2 \div 5$. That might look like a strange division: you have 2 apples, and you are sharing them equally between 5 people. Everybody will get less than 1 apple!

What is the **answer to $2 \div 5$?** It is $\frac{2}{5}$! How can we be sure of that?
Look at the picture. Can you see how 2 pies are divided into five equal parts? You can think of 2 pies shared equally between five people: each person gets two fifths of a pie.

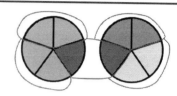

You could also do the division this way: give each person 1 slice of *each* pie.
Again, each person will get two slices, or 2/5 of a pie.

1. Fill in.

a. Divide these 3 pies equally between four

people. Each will get  of a pie.

b. Divide these 2 pies equally between three

people. Each will get 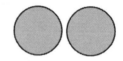 of a pie.

c. Divide these 5 pies equally between six

people. Each will get $\frac{}{}$ of a pie.

d. Divide these 5 pies equally between eight

people. Each will get $\frac{}{}$ of a pie.

e. $\frac{3}{5}$ is the same as the division problem _____ ÷ _____.

If three pizzas are divided equally between five people, each person gets 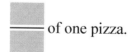 of one pizza.

f. The answer to the division $7 \div 11$ is .

g. $8 \div 21 = \frac{}{}$

h. $21 \div 100 = \frac{}{}$

138

To check a division problem, multiply the answer by the divisor. You should get the dividend.

Example 2. $8 \div 9 = \dfrac{8}{9}$. We can check this division by multiplying $\dfrac{8}{9} \times 9$.

You can either solve this multiplication the regular way: $\dfrac{8}{9} \times 9 = \dfrac{72}{9} = 8$.

Or, you can simplify before you multiply: $\dfrac{8}{\overset{}{\underset{1}{9}}} \times \overset{1}{9} = 8$.

Either way, we get the *original dividend* 8, so the division problem was done correctly.

2. Solve. Check each division by a multiplication.

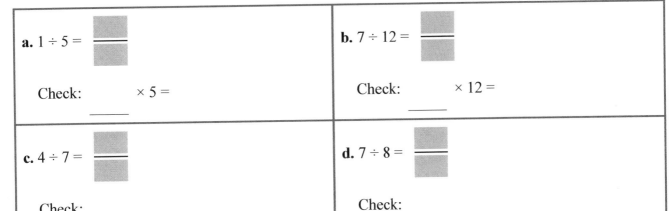

a. $1 \div 5 = \dfrac{}{}$

Check: _____ $\times 5 =$

b. $7 \div 12 = \dfrac{}{}$

Check: _____ $\times 12 =$

c. $4 \div 7 = \dfrac{}{}$

Check:

d. $7 \div 8 = \dfrac{}{}$

Check:

The same idea works with improper fractions: an improper fraction is also a division problem.

Example 3. The answer to $11 \div 4$ is $\dfrac{11}{4}$. Of course, we can write that as the mixed number 2 ¾.

So if four people share 11 apples equally, each person gets 2 ¾ apples.

3. Fill in.

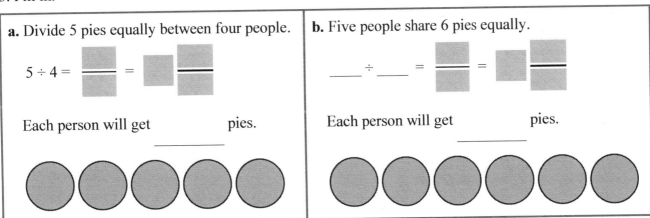

a. Divide 5 pies equally between four people.

$5 \div 4 = \dfrac{}{} = \square \dfrac{}{}$

Each person will get _____ pies.

b. Five people share 6 pies equally.

_____ $\div$ _____ $= \dfrac{}{} = \square \dfrac{}{}$

Each person will get _____ pies.

4. Fill in.

a. Divide 11 pies equally between eight people. _____ ÷ _____ =

Each person will get _____ pies.

b. $\dfrac{17}{6}$ is the same as the division problem _____ ÷ _____.

If 17 pizzas are divided equally between six people, each person gets _____ pizzas.

c. $21 \div 8 = \dfrac{}{} = \square\dfrac{}{}$

d. $46 \div 5 = \dfrac{}{} = \square\dfrac{}{}$

But wait! Isn't the answer to 11 ÷ 4 also 2 R3?

Yes, it is. It is a different way of writing the answer. There are, in fact, several ways to give the answer to 11 ÷ 4: it is 2 R3 or 2 ¾ — and you can also give the latter answer as 2.75.

5. Rewrite the divisions with remainders as divisions where the answer is given as a mixed number.

a. $25 \div 8 = 3\ R1$ $\dfrac{25}{8} = 3\dfrac{1}{8}$	**b.** $44 \div 5 = 8\ R4$ $\dfrac{}{} = \square\dfrac{}{}$	**c.** $23 \div 2 = $ ___ R ___ $\dfrac{}{} = \square\dfrac{}{}$
d. $28 \div 3 = $ ___ R ___ $\dfrac{}{} = \square\dfrac{}{}$	**e.** $65 \div 10 = $ ___ R ___ $\dfrac{}{} = \square\dfrac{}{}$	**f.** $53 \div 9 = $ ___ R ___ $\dfrac{}{} = \square\dfrac{}{}$

6. **a.** Between what two whole numbers is the answer to 45 ÷ 6?

b. You divide five chocolate bars equally between you and your two sisters. How much does each person get?

Study carefully examples 4 and 5 below. The way you give the answer to a division problem depends on the *type* of problem. Some things cannot be divided into parts, so you need to give the answer with the remainder. Some things can be divided into parts, so you give the answer as a mixed number.

Example 4. A teacher shares 65 pencils equally with 26 students. How many will each get?

It does not make sense to break a pencil into parts, so we will give the answer with a remainder: $65 \div 26 = 2$ R13. Each student gets two pencils, and 13 are left over.

Example 5. Twenty-six students share 65 apples equally. How many apples will each get?

$65 \div 26 = \dfrac{65}{26} = 2\dfrac{13}{26} = 2\dfrac{1}{2}$. Each student gets 2 ½ apples.

7. Alison bagged 15 lb of berries equally into four bags.
   How many pounds of berries did she put into one bag?

8. Seventy-five people are organized into groups of 4, as evenly as possible.
   How many and what kind of groups do they get?

9. Calculate

a. 1/12 of 15 kg	**b.** 1/4 of 7 inches

10. If 9 people want to share a 50-pound sack of rice equally by weight, how many pounds of rice should each person get?

    Between what two whole numbers does your answer lie?

11. One mini-bus can seat 11 people. How many mini-buses do you need to transport 102 people?

12. Noah needs to pour 5 liters of juice evenly into 20 glasses.

    a. How many liters is in one glass?

    b. How many milliliters is that?

Dividing Fractions 1: Sharing Divisions

First, let's **share pieces** of pie **evenly among a certain number of people**.
This means that we divide a <u>fraction by a whole number</u>.

$\frac{4}{5}$ of a pie is divided between two people. Each person gets $\frac{2}{5}$ of the pie. $\frac{4}{5} \div 2 = \frac{2}{5}$ Check: $\frac{2}{5} \times 2 = \frac{4}{5}$	$\frac{9}{10}$ is divided between three people. Each person gets $\frac{3}{10}$ of the pie. $\frac{9}{10} \div 3 = \frac{3}{10}$ Check: $\frac{3}{10} \times 3 = \frac{9}{10}$

Notice how we can check each division by multiplication!

1. Color each person's share with a different color, and write a division sentence.

a. $\frac{4}{6}$ of a pie is divided between four people.	**b.** $\frac{3}{5}$ of a pie is divided between three people.
c. $\frac{6}{9}$ of a pie is divided between two people.	**d.** $\frac{6}{10}$ of a pie is divided between three people.
e. $\frac{6}{12}$ of a pie is divided between three people.	**f.** $\frac{15}{20}$ of a pie is divided between five people. 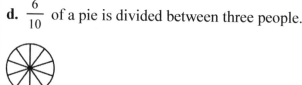

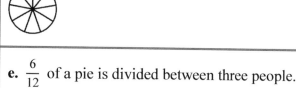

2. Write a division sentence for each problem, and solve it.

a. There is 6/9 of a pizza left over, and three people share it equally. How much does each one get?	**b.** A cake was cut into 20 pieces, and now there are 12 pieces left. Four people share those equally. What fraction of the original cake does each person get?

Next, we divide unit fractions—fractions such as 1/2, 1/3, 1/5, 1/8, 1/12, *etc.* (of the form 1/*n*).

One-half is divided equally between four people. Each person gets 1/8 of it. Can you see why?	One-fifth is divided among three people. Each person gets 1/15. To see that, cut each fifth (colored and uncolored) into three new parts.
$\frac{1}{2} \div 4 = \frac{1}{8}$ Check: $\frac{1}{8} \times 4 = \frac{4}{8} = \frac{1}{2}$	$\frac{1}{5} \div 3 = \frac{1}{15}$ Check: $\frac{1}{15} \times 3 = \frac{3}{15} = \frac{1}{5}$

3. Split the unit fraction equally between the people. Write a division sentence. Write a multiplication sentence to check your division.

a. Divide between two people.	**b.** Divide between two people.	**c.** Divide between two people.
$\frac{1}{2} \div 2 = $ 	$\frac{1}{3} \div 2 = $	$\frac{1}{5} \div 2 = $
Check:  $\times 2 = $	Check: $\times 2 = $	Check: $\times$ ___ $ = $
d. Divide between two people.	**e.** Divide between five people.	**f.** Divide between four people.
$\div 2 = $	 $\div 5 = $	
g. Divide between four people.	**h.** Divide between three people.	**i.** Divide between three people.

Here is a shortcut for **dividing a unit fraction 1/*n* by a whole number *m*:** $\frac{1}{n} \div m = \frac{1}{m \times n}$.

Example. $\frac{1}{8} \div 7 = \frac{1}{56}$. Multiply the denominator of the unit fraction by the divisor to get the new denominator.

4. Solve.

a. $\frac{1}{6} \div 2 = $	**b.** $\frac{1}{10} \div 2 = $	**c.** $\frac{1}{7} \div 3 = $	**d.** $\frac{1}{8} \div 5 = $
e. $\frac{12}{20} \div 2 = $	**f.** $\frac{1}{2} \div 14 = $	**g.** $\frac{8}{5} \div 4 = $	**h.** $\frac{1}{9} \div 9 = $

5. Three children share 1/4 lb of chocolate equally.

   a. How much does each one get, in pounds?

   b. In ounces?

6. A half liter of juice is poured evenly into five glasses.

   a. How much juice is in each glass, measured in liters?

   b. How many milliliters of juice is in each glass?

7. There are 12 beakers with various amounts of oil in them. The line plot shows how much oil each beaker has, in cups.

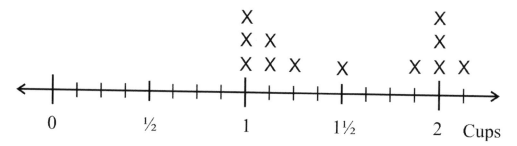

   If all the oil in the beakers was poured together, and then distributed evenly into the 12 beakers, how much oil would be in each beaker?

8. Solve.

a. $\dfrac{2}{9} \div 2 =$	**b.** $\dfrac{1}{9} \div 2 =$	**c.** $\dfrac{14}{20} \div 7 =$	**d.** $\dfrac{8}{11} \div 4 =$
e. $\dfrac{6}{9} \div 3 =$	**f.** $\dfrac{8}{15} \div 4 =$	**g.** $\dfrac{21}{100} \div 3 =$	**h.** $\dfrac{1}{11} \div 2 =$

9. Solve the opposite problem: if each person got this much pie, how much was there originally?

a. $\dfrac{}{} \div 3 = \dfrac{1}{4}$	**b.** $\dfrac{}{} \div 2 = \dfrac{2}{5}$	**c.** $\dfrac{}{} \div 6 = \dfrac{1}{7}$	**d.** $\dfrac{}{} \div 3 = \dfrac{3}{10}$

10. Write a story problem to match each division, and solve.

a. $\dfrac{1}{2} \div 3 =$

b. $\dfrac{6}{8} \div 2 =$

c. $\dfrac{1}{4} \div 2 =$

11. One morning, Joshua's gas can was only 1/8 full.
He poured half of it into his lawn mower.

 a. How full is the gas can now?

 b. If the container holds 3 gallons, what is the amount of gasoline left, in gallons?

 (Challenge) How much is left in quarts?

Lastly, we will divide multiple leftover pie pieces between a certain number of people. This is a bit trickier, but I think you can do it!

When 3/4 is divided equally between two people, one fourth piece must be split into two. Each person gets 1/4 and 1/8.	$\dfrac{3}{4} \div 2 = \left[\dfrac{1}{4} + \dfrac{1}{8}\right] = \dfrac{3}{8}$
Another way of solving the same problem is to split *each* fourth piece into 2. This means we first change the 3/4 into 6/8, and then we can divide evenly by 2.	$\dfrac{3}{4} \div 2$ $\downarrow$ $\dfrac{6}{8} \div 2 = \dfrac{3}{8}$

12. The leftover pie is divided equally. How much does each person get? Write a division sentence.

a. Divide 5/6 between two people. First, split each piece into 2 new ones.	b. Divide 2/3 between three people. First, split each piece into 3 new ones.
c. Divide 2/3 between four people. 	d. Divide 3/4 between four people.
e. Divide 2/5 between three people. First, split each piece into 3.	f. Divide 4/5 between three people.

Dividing Fractions 2: Fitting the Divisor

How many times does one number go into another? You can always write a **division** from this situation. It is always the **divisor that goes or fits into the dividend**.

How many times does go into ? Eight times. We can write the division $2 \div \frac{1}{4} = 8$. Then check the division: $8 \times \frac{1}{4} = \frac{8}{4} = 2$.	How many times does $\frac{1}{2}$ go into 3? Six times. We can write the division $3 \div \frac{1}{2} = 6$. Then check the division: $6 \times \frac{1}{2} = \frac{6}{2} = 3$.

1. Solve. Write a division. Then write a multiplication that checks your division.

a. How many times does 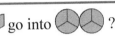 go into 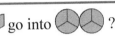 ? $2 \div \frac{1}{3} = $ _____ Check: _____ $\times \frac{1}{3} = $	**b.** How many times does go into ? $1 \div \frac{1}{4} = $ _____ Check: _____ $\times \frac{1}{4} = $
c. How many times does go into ? $6 \div \frac{1}{3} = $ ____ Check:	**d.** How many times does go into ? $5 \div \frac{1}{4} = $ ____ Check:

Now you write the division. Be careful: the divisor is the number that "goes into" the dividend.

e. How many times does  go into  ? ___ $\div$  $=$ Check:	**f.** How many times does  go into  ? ___ $\div$ ___ $=$ Check:
g. How many times does $\frac{1}{6}$ go into 2? ___ $\div$ ___ $=$	**h.** How many times does $\frac{1}{5}$ go into 3? ___ $\div$ ___ $=$

2. Divide. Think, "How many times does the *divisor* go into the *dividend*?" Use the pictures to help.

 **a.** $3 \div \dfrac{1}{6} =$	**b.** $4 \div \dfrac{1}{9} =$	**c.** $4 \div \dfrac{1}{8} =$
d. $3 \div \dfrac{1}{2} =$	**e.** $3 \div \dfrac{1}{7} =$ **f.** $4 \div \dfrac{1}{5} =$	**g.** $2 \div \dfrac{1}{3} =$

Did you notice a pattern? There is a **shortcut** for dividing a whole number by a unit fraction!

$$5 \div \frac{1}{4}$$
$$\downarrow \quad \downarrow$$
$$5 \times 4 = 20$$

$$3 \div \frac{1}{8}$$
$$\downarrow \quad \downarrow$$
$$3 \times 8 = 24$$

$$9 \div \frac{1}{7}$$
$$\downarrow \quad \downarrow$$
$$9 \times 7 = 63$$

Why does it work that way? For example, consider the problem $5 \div (1/4)$. Since 1/4 goes into 1 exactly 4 times, it must go into 5 exactly $5 \times 4 = 20$ times.

3. Solve. Use the shortcut.

a. $3 \div \dfrac{1}{6} =$	**b.** $4 \div \dfrac{1}{5} =$	**c.** $3 \div \dfrac{1}{10} =$	**d.** $5 \div \dfrac{1}{10} =$
e. $7 \div \dfrac{1}{4} =$	**f.** $4 \div \dfrac{1}{8} =$	**g.** $4 \div \dfrac{1}{10} =$	**h.** $9 \div \dfrac{1}{8} =$

4. Write a division for each word problem, and solve. Do *not* write just the answer.

a. How many 1/2-meter pieces can you cut from a roll of string that is 6 meters long?

b. How many 1/4-cup servings can you get from 2 cups of almonds?

c. Ben has small weights that weigh 1/10 kg each. How many of those would he need to make 5 kg?

d. An eraser is 1/8 inches thick. How many erasers can be stacked into a 4-inch tall box?

5. Write a story problem to match each division, and solve.

   a. $2 \div \dfrac{1}{2} =$

   b. $5 \div \dfrac{1}{3} =$

6. These divisions are not as easy as the previous ones, but they are not difficult either. Again, think how many times the divisor goes into the dividend. The pictures can help.

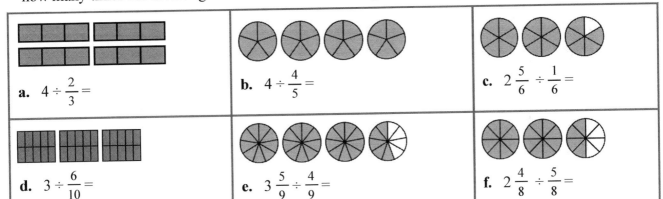

   a. $4 \div \dfrac{2}{3} =$

   b. $4 \div \dfrac{4}{5} =$

   c. $2\dfrac{5}{6} \div \dfrac{1}{6} =$

   d. $3 \div \dfrac{6}{10} =$

   e. $3\dfrac{5}{9} \div \dfrac{4}{9} =$

   f. $2\dfrac{4}{8} \div \dfrac{5}{8} =$

7. Write a division and solve. Write also a multiplication to check your division.

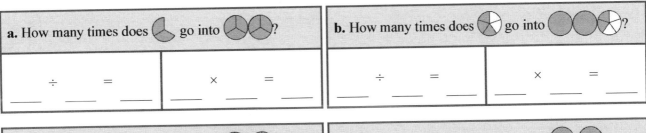

   a. How many times does ⬖ go into ◓◓?

   $\underline{\quad} \div \underline{\quad} = \underline{\quad}$ $\underline{\quad} \times \underline{\quad} = \underline{\quad}$

   b. How many times does ◔ go into ◯◯◔?

   $\underline{\quad} \div \underline{\quad} = \underline{\quad}$ $\underline{\quad} \times \underline{\quad} = \underline{\quad}$

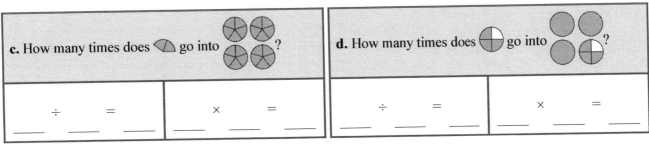

   c. How many times does ◹ go into ◓◓◓◓?

   $\underline{\quad} \div \underline{\quad} = \underline{\quad}$ $\underline{\quad} \times \underline{\quad} = \underline{\quad}$

   d. How many times does ◓ go into ◯◯◯◔?

   $\underline{\quad} \div \underline{\quad} = \underline{\quad}$ $\underline{\quad} \times \underline{\quad} = \underline{\quad}$

8. A recipe calls for 1/2 cup of butter, among other ingredients.
   Alison had plenty of all of the other ingredients except the butter.
   How many batches of the recipe can she make if she has ...

   a. 3 cups of butter?

   b. 2 ½ cups of butter?

9. Jackie made three apple pies and cut them into twelfths.
   She plans on serving two slices to each guest. How many
   servings will she get out of the three pies?
   Hint: Draw a picture.

10. How many 2/10-liter servings do you get from 1 liter of juice?

    From 4 liters of juice?

11. When Natalie goes jogging, she jogs for 1/4 mile, then walks for 1/4 mile,
    then again jogs for 1/4 mile, and so on. How many 1/4 mile stretches are
    there for her in a jogging track that is 2 1/2 miles long?

12. Jill makes bead necklaces that must be exactly 24 inches long. She has size
    SS beads, which are 1/8-inch thick, and size S beads, which are 1/4-inch thick.

    a. How many beads would be in a necklace
       made solely of SS beads?

Bead	Width
SS	1/8 in
S	1/4 in

    b. How many beads would be in a necklace
       made solely of S beads?

    c. She also makes a necklace with the pattern SS-S-SS-S.
       How many of each kind of bead does she need?

Introduction to Ratios

A **ratio** is simply a **comparison of two numbers** or other quantities.

In the picture on the right, there are three hearts and five stars.
To compare the hearts to the stars, we say that the <u>ratio of hearts to stars</u> is 3:5 (read "three to five").

The ratio of <u>stars to hearts</u> is therefore 5:3 (read "five to three").
Notice that the order in which you mention the members of the ratio matters.

We can use fractions, too, to describe the same picture, like this:

$\frac{3}{8}$ of the shapes are hearts. $\frac{5}{8}$ of the shapes are stars.

You can also compare one quantity to the whole group using a ratio.

The ratio of circles to *all* shapes is 2:9.

This is really similar to saying that $\frac{2}{9}$ of the shapes are circles.

1. Describe the images using ratios and fractions.

a.

The ratio of circles to pentagons is _____ : _____

The ratio of pentagons to all shapes is _____ : _____

 of the shapes are pentagons.

b.

The ratio of diamonds to triangles is _____ : _____

The ratio of triangles to all shapes is _____ : _____

 of the shapes are diamonds.

2. Write the ratios.

   a. The ratio of men to women

   b. The ratio of men to all people

   c. The ratio of women to all people

3. **a.** Draw a picture that represents the ratio 3 small balls : 8 bigger balls.

   b. What is the ratio of bigger balls to all balls?

   c. What is the fraction of bigger balls to all balls?

4. "Translate" between a picture, "fraction language," and "ratio language."

picture:	$\frac{3}{4}$ of the shapes are circles. ▭/▭ of the shapes are triangles.	The ratio of circles to triangles is _____ : _____ .

5. The picture on the right shows some men and some women.

    a. What is the ratio of men to women?

    b. Suppose that each symbol *represents* 6 people.

    How many men are there? _____ How many women? _____

    How many people in all? _____

In this picture, **for every heart there are three stars**. That is one way of expressing a ratio. Here are some other ways:

- The shapes are in a ratio of 1 heart to 3 stars.
- There are three stars for every heart.
- The shapes are in a ratio of 3 stars to 1 heart
- The stars and hearts are in a ratio of 1:3.

Now, maybe you say the ratio of hearts to stars is 5:15, because there are 5 hearts and 15 stars.

That ratio is *also* correct! The ratio of hearts to stars is *both* 5:15 and 1:3. The latter is a **simplified ratio**. Ratios are simplified in the same manner as fractions. You will study this more next year.

6. Fill in.

    a. There are _____ hearts to every _____ stars.

    The ratio of hearts to stars is _____ : _____ or _____ : _____ .

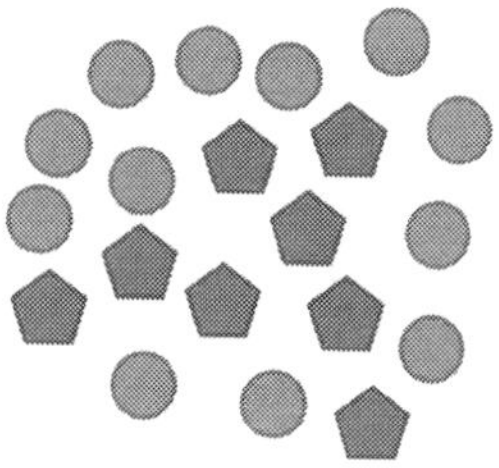

    b. The ratio of pentagons to circles is _____ : _____ or _____ : _____ .

    There are _____ circles to every _____ pentagons.

    c. The ratio of diamonds to triangles is _____ : _____ or _____ : _____ .

    There are _____ diamonds to every _____ triangles.

A theater has 110 plastic chairs in stacks. Three-fifths of the chairs are blue, and the rest are white. We can draw a bar model from this:

Looking at the "blocks" in the model, we can also write a ratio: the ratio of blue chairs to white chairs is 3:2.

Notice: The ratio 3:2 does NOT mention that the chairs are divided into five equal parts, but you can easily see that from the model. (Note that 3 + 2 = 5.)

7. Fill in.

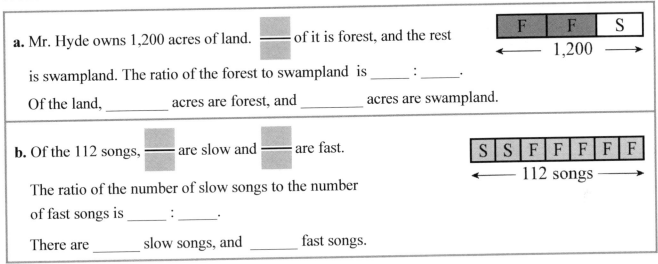

a. Mr. Hyde owns 1,200 acres of land. ____ of it is forest, and the rest

is swampland. The ratio of the forest to swampland is _____ : _____.

Of the land, _____ acres are forest, and _____ acres are swampland.

b. Of the 112 songs, ____ are slow and ____ are fast.

The ratio of the number of slow songs to the number

of fast songs is _____ : _____.

There are _____ slow songs, and _____ fast songs.

Draw a bar model for each situation, and solve it.

8. The ratio of puppies to adult dogs in a kennel is 4:1.
   (Hint: Draw a model with 4 "puppy" blocks and one "adult" block.)

   a. If there are 80 dogs in total, how many dogs
   does each block represent?

   b. How many of the 80 dogs are adults?

   c. How many are puppies?

9. Marsha has 147 marbles in her bag, some white and
   some red. The ratio of white marbles to red ones is 3:4.
   How many marbles are red?

10. Anita and Shirley shared a reward of $200 in a ratio of 3:5.

   a. How much did Anita get?

   b. How much did Shirley get?

11. There are 1,404 students in the local community college. The ratio of male students to female students is 4:5.

   a. Draw a model to represent the situation.

   b. What fractional part of the students are male?

   c. How many male students are there?

12. A paddock contains 102 horses. Five-sixths of the horses are white, and the rest are pintos.

   a. Draw a model to represent the situation.

   b. What is the ratio of white horses to pinto horses?

   c. How many white horses are there?

13. Jack has a box filled with white and transparent marbles in a ratio of 2:5. He has 38 white marbles.

   a. Draw a model to represent the situation.

   b. What fractional part of the marbles are white?

   c. How many marbles does he have in all?

Mixed Review

1. Subtract.

a. $4\frac{2}{6} - 1\frac{5}{6} =$	**b.** $3\frac{2}{9} - 1\frac{7}{9} =$
c. $4\frac{2}{3} - 2\frac{1}{4} =$	**d.** $7\frac{1}{6} - 1\frac{3}{5} =$

2. Write the numbers in expanded form.

   a. 0.28

   b. 60.068

3. There were 780 people at a concert. One-third of them came with a discount ticket. Another 120 were seniors and came with a special low-price ticket. The rest paid a regular-priced ticket. How many people paid the normal price? *(Draw a bar model to help.)*

4. Make a line graph of the baby's weight.

Week	Weight	Weight in ounces
0	6 lb 14 oz	
1	6 lb 12 oz	
2	6 lb 14 oz	
3	7 lb	
4	7 lb 2 oz	
5	7 lb 4 oz	
6	7 lb 6 oz	
7	7 lb 7 oz	

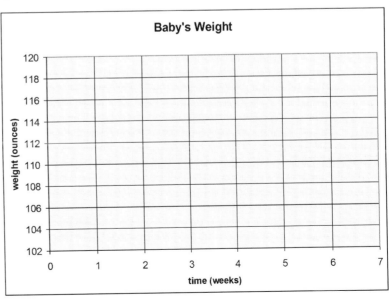

Baby's Weight

155

Plot the points from the "number rules" or number patterns on the coordinate grids.

5. The rule for *x*-values: start at 2, and add 1 each time.
   The rule for *y*-values: start at 0, and add 2 each time.

x	2					
y	0					

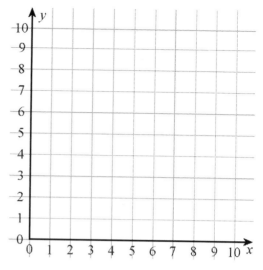

6. The rule for *x*-values: start at 1, and add 1 each time.
   The rule for *y*-values: start at 10, and subtract 2 each time.

x						
y						

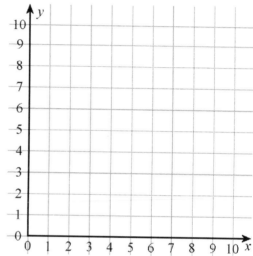

7. Use rounded numbers to estimate the answer.
   How many $0.58 cans can you get with $7?

8. Jack cut three 0.82-meter pieces from a 4-meter board.
   How long was the piece that was left?

9. Factor the following composite numbers to their prime factors.

a. 28 /\	**b.** 55 /\	**c.** 84 /\

10. The data below give the responses of 20 fifth graders to the question, "What is your favorite pet?"

   dog, dog, hamster, dog, cat, cat, parrot, dog, horse, dog, cat, canary, goldfish, dog, cat, cat, dog, dog, canary, hamster

   a. Find the mode.

   b. If possible, calculate the mean.

11. Jenny buys two computer keyboards that had originally cost $15.60 but now they are 2/10 off of their normal price. Find Jenny's total bill.

12. Convert. One mile is 5,280 ft. Round your answers to whole feet.

   a. 0.6 mi = _____ ft

   b. 3.45 mi = _____ ft

13. Match the two problems (a) and (b) below to the correct expressions below them. Then solve each problem. What does your answer signify?

   a. John has $170 to buy groceries for the week. First, John sets aside $23 to buy treats; then he divides the remaining money evenly for each day of the week.

   b. John has $170 to buy groceries for the week. John decides to use $23 per day for food, and to use whatever is left for treats.

   $170 − 7 × $23 $170 − $23 ÷ 7   ($170 − $23) ÷ 7

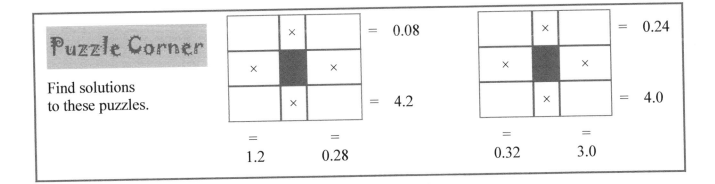

Puzzle Corner

Find solutions to these puzzles.

Chapter 7 Review

1. Simplify the fractions. Complete the pie pictures of the process.

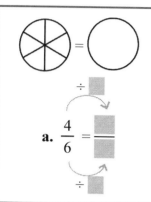

a. $\dfrac{4}{6} = \dfrac{}{}$

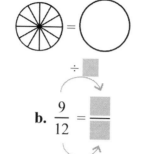

b. $\dfrac{9}{12} = \dfrac{}{}$

c. $\dfrac{24}{30} =$

d. $3\dfrac{15}{35} =$

e. $\dfrac{56}{49} =$

f. $\dfrac{12}{100} =$

g. $\dfrac{45}{27} =$

h. $2\dfrac{72}{84} =$

2. Draw a picture to illustrate these calculations, and solve.

a. $3 \times 1\dfrac{1}{3}$

b. $2 \times \dfrac{5}{6}$

3. Multiply.

a. $7 \times \dfrac{2}{5}$

b. $\dfrac{2}{7} \times \dfrac{5}{6}$

c. $4\dfrac{3}{10} \times 4$

d. $1\dfrac{1}{6} \times 5\dfrac{2}{3}$

4. Simplify before you multiply.

a. $\dfrac{7}{14} \times \dfrac{3}{12}$	**b.** $\dfrac{5}{24} \times \dfrac{12}{30}$

5. Figure out the side lengths of the colored rectangle from the picture. Then multiply the side lengths to find its area. <u>Check that the area you get by multiplying is the same as what you can see</u> from the picture.

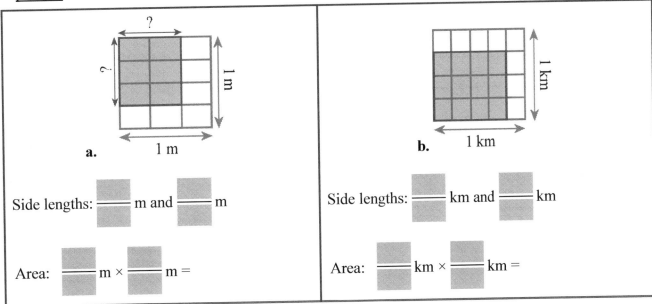

a.

Side lengths: _____ m and _____ m

Area: _____ m × _____ m =

b.

Side lengths: _____ km and _____ km

Area: _____ km × _____ km =

6. Shade a rectangle inside the square so that its area can be found by the fraction multiplication.

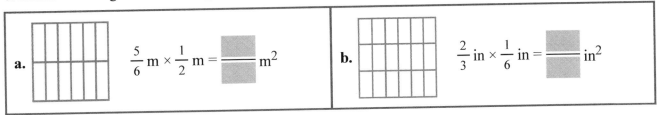

a. $\dfrac{5}{6}$ m × $\dfrac{1}{2}$ m = _____ m^2

b. $\dfrac{2}{3}$ in × $\dfrac{1}{6}$ in = _____ in^2

7. Mary jogs 3/4 of a mile each day, five days a week.
   Calculate how many miles she jogs in a week.

8. Sally made a rectangular blueberry pie and cut it into 20 equal pieces. The next morning, 12/20 of it was left. Then, the dog got on the table and gobbled up 2/3 of what was left!

   a. How many pieces are left now?

   b. What fraction of the pie is left now?

9. Draw a picture to illustrate these calculations, and solve.

a. $1 \div 3$	**b.** $\dfrac{1}{2} \div 3$

10. Divide.

a. $2 \div \dfrac{1}{3}$	**b.** $4 \div \dfrac{1}{4}$	**c.** $\dfrac{1}{2} \div 5$
d. $\dfrac{1}{7} \div 3$	**e.** $9 \div 4$	**f.** $\dfrac{1}{8} \div 2$
g. $6 \div 9$	**h.** $\dfrac{6}{10} \div 3$	**i.** $\dfrac{3}{4} \div 3$

11. Solve.

a. A string that is 7 inches long is cut into four equal pieces.
How long are the pieces?

b. Find four-fifths of the fraction 1/3.

c. Five people are sharing equally 11 lb of almonds.
How many pounds will each get?

d. Chain costs $24 per meter, and you bought 3/4 of a meter.
What was the cost?

e. There were 112 contestants, and 3/8 of them were women.
How many were not?

160

12. Is the result of multiplication more, less, or equal to the original number? You do not have to calculate anything. Compare and write <, >, or = in the box.

a. $\frac{8}{9} \times 7$ ☐ 7	b. $2\frac{1}{11} \times 57$ ☐ 57	c. $\frac{7}{7} \times 13$ ☐ 13

13. One-third of a cake was decorated with chocolates, one-fourth with sprinkles, and the rest with strawberry frosting. What *part* was decorated with strawberry frosting?

14. A loaf of bread was cut into 30 slices. After a day, 5/6 of it was left. Then the family ate 1/5 of the *remaining* bread. How many slices are left now?

15. You only have 3/4 cup of walnuts in the cupboard, so you decide to make only 3/4 of the recipe. How much of each ingredient do you need?

Brownies

3 cups sweetened carob chips
8 tablespoons olive oil
2 eggs
1/2 cup honey
1 teaspoon vanilla
3/4 cup whole wheat flour
3/4 teaspoon baking powder
1 cup walnuts or other nuts

16. **a.** Determine which sheet of paper has the greater area:
(1) a 6 ½ in by 8½ in sheet or (2) a 5¾ in by 9 in sheet.

b. How much greater is the area of the larger sheet than the area of the smaller sheet, in square inches?

Chapter 8: Geometry
Introduction

This chapter includes many problems that involve drawing a geometric figure, because drawing is an excellent help towards achieving a conceptual understanding of geometry. Most of those are marked with the symbol " ", which means the exercise is to be done in a notebook or on blank paper.

The chapter starts out with lessons that review topics from previous grades, such as measuring angles, the vocabulary of basic shapes, how to draw a perpendicular line through a given point on a line, and how to draw a triangle with given angle measurements. Some fun is included, too, with star polygons.

In the lesson about circles, we learn the terms circle, radius, and diameter. Students are introduced to a compass, and they draw circles and circle designs using a compass.

Then we go on to classify quadrilaterals and learn the seven different terms used for them. The focus is on understanding the classification, and understanding that attributes defining a certain quadrilateral also belong to all the "children" (subcategories) of that type of quadrilateral. For example, squares are also rhombi, because they have four congruent sides (the defining attribute of a rhombus).

Next, we study and classify different triangles. Students are now able to classify triangles both in terms of their sides and also in terms of their angles. The lesson also includes several drawing problems where students draw triangles that satisfy the given information.

The last focus of this chapter is volume. Students learn that a cube with the side length of 1 unit, called a "unit cube," is said to have "one cubic unit" of volume, and can be used to measure volume. They find the volume of right rectangular prisms by "packing" them with unit cubes and by using formulas. They recognize volume as additive and solve both geometric and real-word problems involving volume of right rectangular prisms.

The Lessons in Chapter 8

Helpful Resources on the Internet

FOR REVIEW OF ANGLES AND POLYGONS

Measuring Angles
Rotate the protractor into position and give your measurement to the nearest whole number.
http://www.mathplayground.com/measuringangles.html

Turtle Pond
Guide a turtle to a pond using commands that include turning him through certain angles and moving him specific distances.
http://illuminations.nctm.org/Activity.aspx?id=3534

Interactive Polygon Crossword Puzzle
Use the clues to help you guess the words that go in the puzzle, and fill it in.
http://www.mathgoodies.com/puzzles/crosswords/ipolygon3.html

Types of Polygons Vocabulary Quiz
In this interactive quiz you have to quickly name different types of polygons based on given clues. For each question you will have only 30 seconds to write your answer!
http://www.math-play.com/types-of-poligons.html

Polygon Matching Game
Many of the polygons included are quadrilaterals.
http://www.mathplayground.com/matching_shapes.html

Free Worksheets for Area and Perimeter
Create worksheets for the area and the perimeter of rectangles/squares with images, word problems, or problems where the student writes an expression for the area using the distributive property. Options also include area and perimeter problems for irregular rectangular areas, and more.
http://www.homeschoolmath.net/worksheets/area_perimeter_rectangles.php

Areas of Rectangular Shapes Quiz
Practice finding the area of rectangular compound shapes with this interactive quiz.
https://www.studyladder.com/games/activity/area-of-irregular-shapes-13136

Circle
This page includes a detailed lesson about circles, as well as interactive exercises to practice the topic.
http://www.mathgoodies.com/lessons/vol2/geometry.html

QUADRILATERALS

Interactive Quadrilaterals
See all the different kinds of quadrilateral "in action." You can drag the corners, see how the angles change, and observe what properties do not change.
http://www.mathsisfun.com/geometry/quadrilaterals-interactive.html

Properties of Quadrilaterals
Investigate the properties of a kite, a rhombus, a rectangle, a square, a trapezoid, and a parallelogram in this dynamic, online activity.
https://www.geogebra.org/m/yekC7cDh

Complete the Quadrilateral
This is a hands-on activity (printable worksheets) where students join the dots to complete quadrilaterals, which helps students learn about the different types of quadrilaterals.
http://fawnnguyen.com/don-stewards-complete-quadrilateral/

Types of Quadrilaterals Quiz
Identify the quadrilaterals that are shown in the pictures in this interactive multiple-choice quiz.
http://www.softschools.com/math/geometry/quadrilaterals/types_of_quadrilaterals/

Quadrilateral Types Practice at Khan Academy

Identify quadrilaterals based on pictures or attributes in this interactive quiz.

https://www.khanacademy.org/math/basic-geo/basic-geo-shapes/basic-geo-classifying-shapes/e/quadrilateral_types

Classify Quadrilaterals Worksheets

Make free printable worksheets for classifying (identifying, naming) quadrilaterals.

http://www.homeschoolmath.net/worksheets/classify_quadrilaterals.php

TRIANGLES

Triangle Shoot

Practice classifying triangles by their angles or by their sides, or identifying types of angles, with this "math splat" game.

http://www.sheppardsoftware.com/mathgames/geometry/shapeshoot/triangles_shoot.htm

Rags to Riches: Classify Triangles by Sides and Angles

Answer multiple-choice questions about classifying triangles by their angles and sides and about angle measures of a triangle in a quest for fame and fortune.

http://www.quia.com/rr/457498.html

Identify Triangles Quiz

A simple multiple-choice quiz about identifying (classifying) triangles either by their sides or angles. You can modify some of the quiz parameters, such as the number of problems in it.

http://www.thatquiz.org/tq-A/?-j1-l34-p0

Interactive Triangles Activity

Play with different kinds of triangles (scalene, isosceles, equilateral, right, acute, obtuse). Drag the vertices and see how the triangle's angles and sides change.

https://www.mathsisfun.com/geometry/triangles-interactive.html

Classify Triangles Worksheets

Make free printable worksheets for classifying triangles by their sides, angles, or both.

http://www.homeschoolmath.net/worksheets/classify_triangles.php

VOLUME

Geometric Solids

Rotate various geometric solids by dragging with the mouse. Count the number of faces, edges, and vertices.

http://illuminations.nctm.org/Activity.aspx?id=3521

Cuboid Exploder and Isometric Shape Exploder

These interactive demonstrations let you see either various cuboids (a.k.a. boxes or rectangular prisms) or various shapes made of unit cubes, and then "explode" them to the unit cubes, illustrating volume.

http://www.teacherled.com/resources/cuboidexplode/cuboidexplodeload.html **and**
http://www.teacherled.com/resources/isoexplode/isoexplodeload.html

3-D Boxes Activity

Identify how many cubes are in the 3-D shapes in this interactive activity.

http://www.interactivestuff.org/sums4fun/3dboxes.swf

Rectangular Prisms Interactive Activity

Fill a box with cubes, rows of cubes, or layers of cubes. Can you determine a rule for finding the volume of a box if you know its width, depth, and height?

http://illuminations.nctm.org/Activity.aspx?id=4095

Interactivate: Surface Area and Volume

Explore or calculate the surface area and volume of rectangular prisms and triangular prisms. You can change the base, height, and depth interactively.

http://www.shodor.org/interactivate/activities/SurfaceAreaAndVolume/

Decompose Figures To Find Volume - Practice at Khan Academy
Find the volume of irregular 3-D figures by dividing the figures into rectangular prisms and finding the volume of each part.
https://www.khanacademy.org/math/cc-fifth-grade-math/cc-5th-measurement-topic/cc-5th-volume/e/decompose-figures-to-find-volume

Volume Word Problems
Practice solving word problems that involve volume of rectangular prisms.
https://www.khanacademy.org/math/pre-algebra/measurement/volume-introduction-rectangular/e/volume_2

Worksheets for the Volume and Surface Area of Rectangular Prisms
Customizable worksheets for volume or surface area of cubes and rectangular prisms. Includes the option of using fractional edge lengths.
http://www.homeschoolmath.net/worksheets/volume_surface_area.php

FOR FUN

Patch Tool
An online activity where the student designs a pattern using geometric shapes.
http://illuminations.nctm.org/Activity.aspx?id=3577

Shape Guess - Elimination Game
Have fun with shapes while playing this interactive online guessing game!
http://www.learnalberta.ca/content/mejhm/index.html?
l=0&ID1=AB.MATH.JR.SHAP&ID2=AB.MATH.JR.SHAP.SHAP&lesson=html/object_interactives/shape_classification/use_it.html

Interactivate! Tessellate
An online, interactive tool for creating your own tessellations. Choose a shape, then edit its corners or edges. The program automatically changes the shape so that it will tessellate (tile) the plane. Then push the tessellate button to see your creation! Requires Java.
http://www.shodor.org/interactivate/activities/Tessellate

Review: Angles

An angle is a figure formed by two **rays** that have the same beginning point. That point is called the **vertex**. The two rays are called the sides of the angle.

Imagine the two sides as being like two sticks that open up a certain amount. The more they open, the bigger the angle.

An angle can be named (1) after the vertex point, (2) with a letter inside the angle, or (3) using one point on the ray, the vertex point, and one point on the other ray.

We measure angles in degrees. You can use a protractor like the one at the right to measure angles. The angle in blue measures 35 degrees.

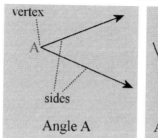

Angle A

Angle α

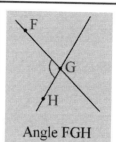

Angle FGH

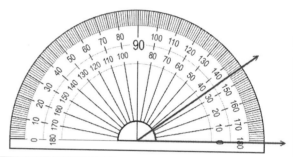

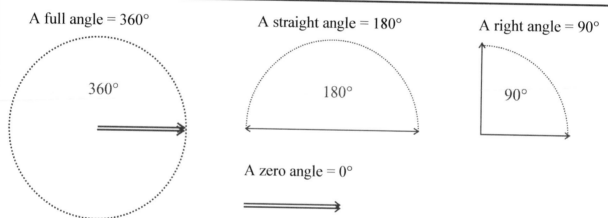

A full angle = 360°

360°

A straight angle = 180°

180°

A right angle = 90°

90°

A zero angle = 0°

Angles that are more than 0° but less than 90° are called **acute** ("sharp") angles.
Angles that are more than 90° but less than 180° are called **obtuse** ("dull") angles.
(Angles that are more than 180° but less than 360° are called *reflex* angles.)

1. Continue the sides of these angles with a ruler.

   Then, measure them with a protractor.

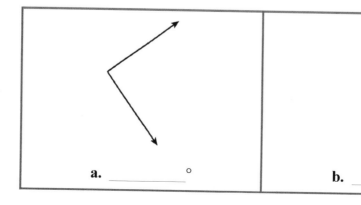

a. _____ °

b. _____ °

2. In your notebook, draw:

  a. Any acute angle. Measure it. Label the angle as "An acute angle, xx°."

  b. Any obtuse angle. Measure it. Label the angle as "An obtuse angle, xx°."

3. Draw three dots on a blank paper and join them to form a triangle.
   Draw the dots far enough apart so that the triangle nearly fills the page.
   Then, measure the angles of your triangle.

   The angles of my triangle are: _____°, _____°, and _____°.

   What is the *sum* of these angle measures? _____°

4. Draw a horizontal line and mark a point on it. This point will be the vertex of an angle.
   Draw the other side of the angle from the vertex so that the angle measures 76°.

5. Follow the procedure above to draw acute angles with the following measures:
   a. 30°   **b.** 60°   **c.** 45°

6. Draw obtuse angles with the following measures:
   a. 135°   **b.** 100°   **c.** 150°

7. Now that you have drawn several angles, *estimate* the angle measure of these angles. Write down the
   estimates on the top lines. Then measure the angles, and write down the measures on the bottom lines.
   To measure the angles, you will need to continue their sides.

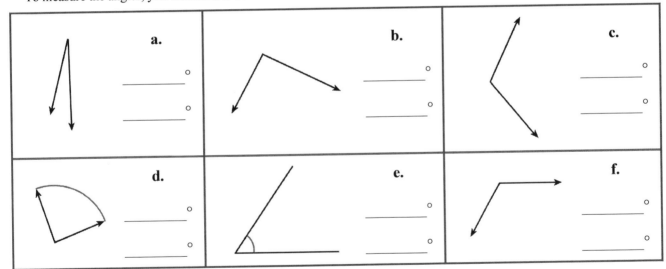

Important Terms
• *an angle*   • *an acute angle*
• *a zero angle*   • *a right angle*
• *a straight angle* • *an obtuse angle*

Review: Drawing Polygons

Review these terms for geometric figures:

- A **polygon** – a closed figure made up of line segments.
- A **right triangle** – a triangle with one right angle.
- An **obtuse triangle** – a triangle with one obtuse angle.
- An **acute triangle** – a triangle with all three angles acute.
- A **quadrilateral** – a polygon with four sides.

- A **pentagon** – a polygon with *five* sides.
- A **hexagon** – a polygon with *six* sides.

- A **heptagon** – a polygon with *seven* sides.
- An **octagon** – a polygon with *eight* sides.

- A **vertex** is a "corner" of a polygon.
- A **diagonal** is a line segment drawn from one vertex of a polygon to another (inside the polygon).

These pictures remind you how to use a protractor or a triangular ruler to draw a **perpendicular** line through a given point (a line that is at a right angle with the given line).

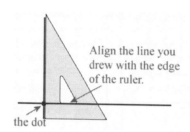

Align the line you drew with the edge of the ruler.

the dot

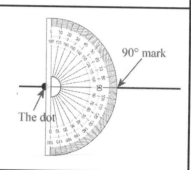

90° mark

The dot

1. Draw any pentagon in your notebook by first drawing five DOTS, and then connecting them with line segments (use a ruler!) Draw the dots kind of randomly around the page and so that your pentagon <u>nearly fills the space</u>. You will need a fairly large pentagon. Don't try to get a regular shape but a pentagon with different side lengths and angles.

   a. Measure all the angles of your pentagon. The angles measure:

   _____ ° , _____ ° , _____ ° , _____ ° , and _____ .

   b. Now draw two diagonals inside your pentagon, dividing it into three triangles. Classify each of those triangles as acute, right, or obtuse.

2. Draw a perpendicular line to the given line through the given point.

   (If you need more practice, repeat this task in your notebook. Start by drawing a line and a point on it.)

   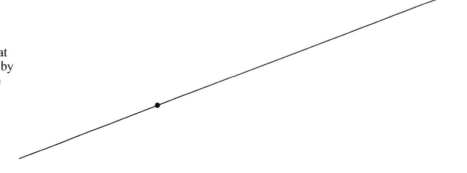

3. Draw a square with 3¼-inch sides. Use a proper tool for drawing perpendicular lines.

(1) Draw a long line, longer than necessary. Mark on it the first 3¼-inch side.

(2) Now draw two perpendicular lines from the two endpoints of the 3¼-inch side.

(3) Measure the other two 3¼-inch sides. Draw dots where the two vertices will be.

(4) Draw in the last side of your square.

How to draw a triangle with two given angle measurements.

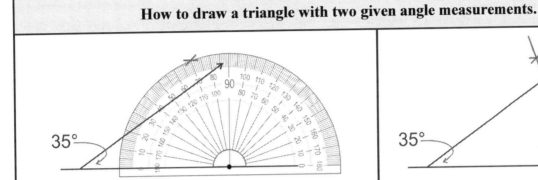

Let's say you have already drawn a 35° angle, and the second angle is supposed to be 70°. The image shows you how to place your protractor so you can measure and mark the 70° angle.

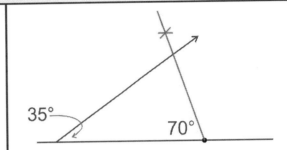

Then remove the protractor and draw the third side of the triangle.

---

4. **a.** Draw a triangle with 50° and 75° angles.

b. Measure the third angle. It measures _____°.

c. What is the sum of the three angle measures?

5. **a.** Draw a triangle with 110° and 35° angles.

b. Measure the third angle. It measures _____°.

c. What is the sum of the three angle measures?

Important Terms		
• a right triangle	• vertex	• pentagon
• an acute triangle	• diagonal	• hexagon
• an obtuse triangle	• perpendicular	• heptagon
• polygon	• quadrilateral	• octagon

Star Polygons (*optional*)

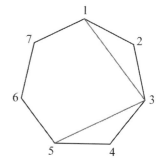

1. This picture shows a regular heptagon where every *other* vertex is connected with a line segment (skipping one vertex in between).

   Continue drawing diagonals in such a manner. The shape you will get is called a *star polygon*, and specifically a *heptagram*.

2. Make star polygons.

a. This is a regular nonagon. Make a nonagram connecting every four vertices (skipping 3).

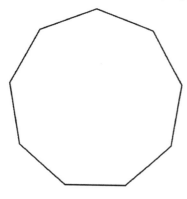

b. This is a regular pentagon. Make a pentagram connecting every other vertex (skipping one).

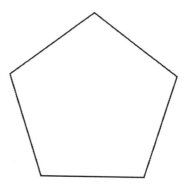

c. This is a regular octagon. Make an octagram connecting every three vertices (skipping 2).

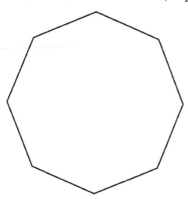

d. This is a regular heptagon. Make a heptagram connecting every three vertices (skipping 2).

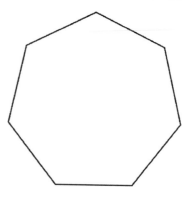

e. This is a regular decagon. Make a decagram connecting every three vertices (skipping 2).

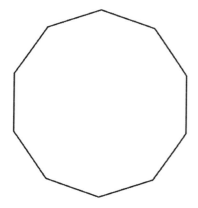

f. Make a nonagram connecting every second vertex (skipping 1).

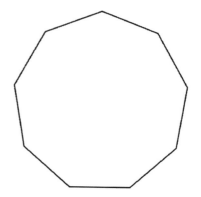

3. You can make your own star polygons here. Experiment!

Circles

These figures are round, but they are not circles. 	These are ovals. They are symmetric and round, but they are still not circles. Why not? What makes a circle?	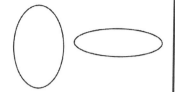

The difference between other round figures and circles is this:

In a circle, the <u>distance</u> from the **center point** to the actual circle line, or **circumference of the circle**, remains the same.

This distance is called the **radius** of the circle.

In other words, all the points on the circumference are **at the same distance from the center point**.

The distance from the center point to any point on the circumference is called the **radius**.	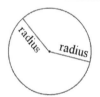	A line through the center point is called a **diameter**.	

1. Draw a radius or a diameter from the given point. Use a ruler. Look at the example.

 Here, a radius is drawn from the given point.	**a.** Draw a radius from the given point. 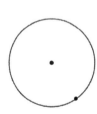	**b.** Draw a radius from each of the given points. 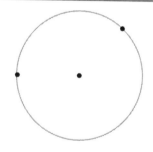
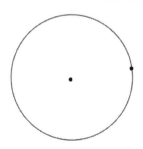 **c.** Draw a diameter from the given point.	**d.** Draw a diameter for the smaller circle and a diameter for the bigger circle from the given points. 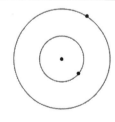	**e.** Draw a radius from the point A and a diameter from the point B. 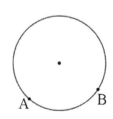

172

2. Learn to use a compass to draw circles.

   a. Draw many circles with the compass.

   b. Now, set the radius on the compass to be 3 cm, and draw a circle.
   You can do that by placing the compass next to a ruler, and adjusting
   the radius of the compass until it is 3 cm as measured by the ruler.
   Some compasses show the radius for you, so you won't need a ruler.

   c. Draw a circle with a radius of 5 cm.

   d. Draw a circle with a radius of 1 ½ in.

3. **a.** Draw two diagonals into this square. Draw a point
   where they cross (the center point of the square).
   Now, erase the lines you drew, leaving the point.

   b. Draw a circle *around* the square so that it touches
   the vertices of the square. Use the point you drew
   in (a) as the center point.

   c. Fill in: The _____ of the circle
   has the same length as the diagonal of the square.

4. **a.** Draw a circle *inside* this square so that it touches
   the sides of the square but will not cross over them.

   b. Fill in: The _____ of the square
   has the same length as the diameter of the circle.

   You can repeat or practice exercises #3 and #4 in
   your notebook.

5. **a.** Draw a circle with center point (5, 6)
   and a radius of 2 units. Use a compass.

   b. Draw another circle with the same center
   point, but double the radius.

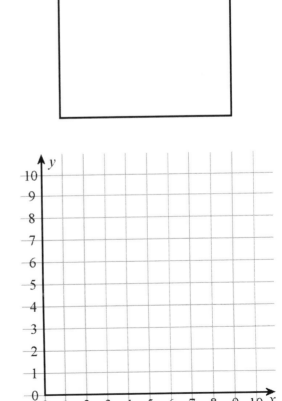

6. Draw these figures using a compass and a ruler in your notebook. Your figures don't have to be the same size as these; they just need to show the same pattern. *See hints at the bottom of this page. Optionally, you can also draw these in drawing software.*

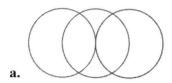

a.

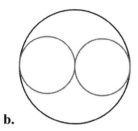

b.

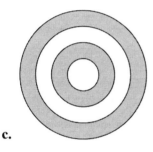

c.

d.

See **http://homeschoolmath.blogspot.com/2013/02/geometric-art-project-seven-circle.html** for one more circle design and art project!

a. Hint: Draw a line. Then, draw the three center points on it, equally spaced.

b. Hint: First, draw the three center points for the three circles, equally spaced. What is the radius of the big circle compared to the radius of the small ones?

c. Hint: What pattern is there in the radii of these circles? These circles are called concentric circles because they share the same center point.

d. Hint: You need to draw the outer square first. Then measure and divide it into quarters. Measure to draw the center points of the circles (they are midpoints of the sides of the smaller squares).

```
New terms to remember:
• circle            • radius
• circumference     • diameter
```

Quadrilaterals

Quadrilaterals are polygons with four sides (*quadri-* = four, *lateral* = referring to a side). You already know about the three quadrilaterals below.

Note: If two sides of a quadrilateral have the same length, they are said to be **congruent**.

1.	2.	3.
A **parallelogram** has two pairs of parallel sides.	A **rectangle** has four right angles.	A **square** is a rectangle with four congruent sides.

We can organize these three quadrilaterals in a **tree diagram** (on the right).

Start "reading" the tree diagram from the top, beginning with the parallelogram. The next figure, the rectangle, is like a "child" to the parallelogram. If the parallelogram is the parent of the family, then its child has the same "family name" because it belongs to the parallelogram family.

Why? Because a rectangle also has two pairs of parallel sides. So it, too, is a parallelogram! Additionally, its angles are right angles, so a rectangle has something *more* than a parallelogram does.

Similarly, a square is like a "child" of the rectangle. The square has the same properties as its "parent" and "grandparent": the square is also a rectangle, and it is also a parallelogram. Additionally, all of its sides are congruent.

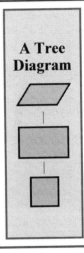

A Tree Diagram

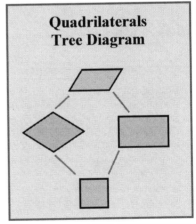

Quadrilaterals Tree Diagram

4.

A **rhombus** is a parallelogram that has four congruent sides (a diamond).

Now let's add a new "family member" to the diagram: the *rhombus*.

The rhombus is a parallelogram, so it belongs under the parallelogram in the tree diagram.

It shares something with the square, as well. Both have four congruent sides. This means the square goes *under* the rhombus. But the rhombus and the rectangle do *not* share characteristics (other than both being quadrilaterals).

1. Draw a quadrilateral that has **four right angles** and **one side 2 inches long**. Can you draw only *one* kind of quadrilateral like that, or can you draw *several* kinds that all look different? (Compare your results with those of your classmates.)

2. A *regular* quadrilateral would be a quadrilateral whose sides are all congruent and whose angles all have the same measure. What is the usual name for such a figure?

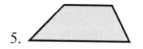 5. A **trapezoid** has *at least* one pair of parallel sides. It may have two!	 6. A **kite** has two pairs of congruent sides that touch each other. *The single tick marks show the one pair of congruent sides, and the double tick marks show the other pair.*	 7. In a **scalene** quadrilateral, all sides are of different lengths (no two sides are congruent).

3. Each quadrilateral below is either a parallelogram, a rhombus, a trapezoid, or a kite. Write their names.

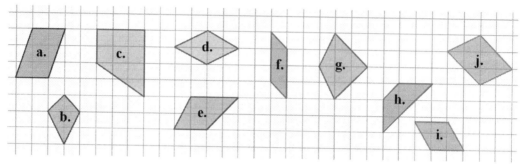

a. _____   b. _____

c. _____   d. _____

e. _____   f. _____

g. _____   h. _____

i. _____   j. _____

4. In the grid below:

    a. Join the dots to make a parallelogram.   **b.** Draw two more different parallelograms.

    c. Draw a rhombus.   **d.** Draw a kite.   **e.** Draw a scalene quadrilateral.

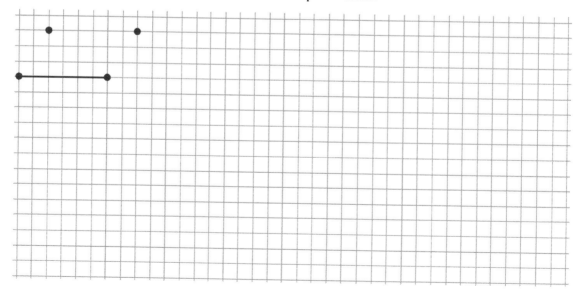

5. Our tree diagram is now complete with seven different kinds of quadrilaterals! Name each type of quadrilateral in the diagram.

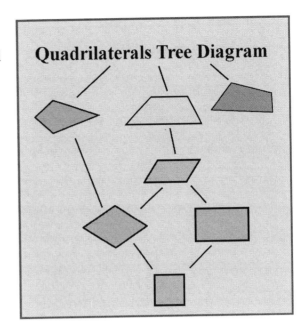

6. Answer the questions. The tree diagram will help.

    a. Is a rhombus also a kite?

    b. Is a square also a kite?

    c. Is a rectangle also a kite?

    d. Is a square also a trapezoid?

    e. Is a parallelogram also a kite?

7. Draw a trapezoid so that its parallel sides measure 3 in and 1 ¾ in. Can you draw several such trapezoids that are not identical?

8. The sides of a parallelogram measure 2 in, 3 in, 2 in, and 3 in. Is it also a kite? A rhombus? A trapezoid?

9. A trapezoid's sides measure 5 cm, 3 cm, 8 cm, and 3 cm. Is it also a kite? A parallelogram? A rhombus?

10. Janine is supposed to draw a quadrilateral with each side 1½ inches long. Can she draw only one kind, or can she draw several different-looking ones? Explain.

11. A certain quadrilateral has two pairs of congruent sides and also two pairs of parallel sides. What kind of quadrilateral is it?

12. Solve the quadrilaterals puzzle and uncover a message!

Quadrilateral	Letter
I have one right angle. I look like a toy that comes back to you.	
I have the biggest area around here!	
I am the little rhombus!	
Two (and only two) of my sides are parallel. The other two are congruent.	
There's nothing congruent or parallel about me.	
I'm the bigger parallelogram with four congruent sides.	
I am the big rhombus!	
I've got right angles and two pairs of congruent sides.	
I have four right angles and four congruent sides!	
Two—and only two—of my sides are parallel. None are congruent.	
They call me a rectangle.	
I'm one of the parallelograms in the bunch with a vowel inside me.	
I have two pairs of parallel sides, but not all of my sides are congruent.	
You can't find a more regular quadrilateral than me.	
While I do have two parallel sides, you cannot draw a line of symmetry through me.	
Trapezoid is my name, but I don't have any congruent sides.	
Scalene—yep, that's me!	
None of my sides are parallel to each other, but I have two pairs of congruent sides.	
I'm the diamond-shape again... the bigger one.	
If you stretch me, I'll be a rectangle.	
Oops! I don't even belong to this group!	
	!

178

Equilateral, Isosceles, and Scalene Triangles

Classification according to sides	
If all three sides of a triangle are congruent (the same length), it is called an **equilateral triangle**. "*Equi-*" refers to things that are the same or equal, and "*lateral*" refers to sides. Think of it as a "same-sided" triangle. 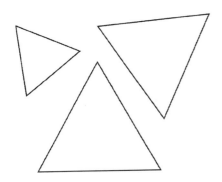	If only *two* of a triangle's sides are congruent, then it is called an **isosceles triangle**. Think of it as a "same-legged" triangle, the "legs" being the two sides that are the same length. MARK the two congruent sides of each isosceles triangle: 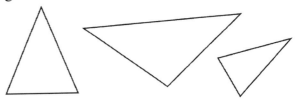
	Lastly, if none of the sides of a triangle are congruent (all are different lengths), it is a **scalene triangle**. 

1. Classify the triangles by the lengths of their sides as either equilateral, isosceles, or scalene.

   You can mark each triangle with an "*e*," "*i*," or "*s*" correspondingly.

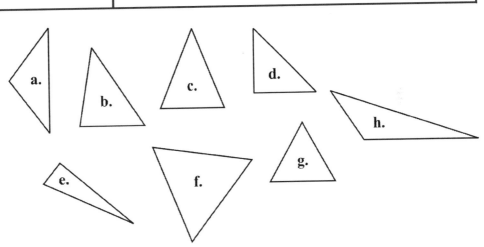

2. Plot the points (0, 0), (3, 5), (0, 5) , and connect them with line segments to form a triangle.

   Classify your triangle by its sides. Is it equilateral, isosceles, or scalene?

   _____

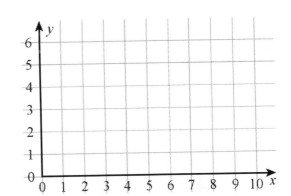

3. Classify the triangles as "acute," "right," or "obtuse" (by their angles), and also as "equilateral," "isosceles," or "scalene" (by their sides).

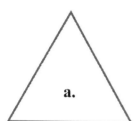
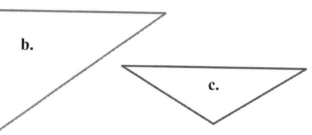

Triangle	Classification by the angles	Classification by the sides
a.		
b.		
c.		
d.		

4. Plot the points, and connect them with line segments to form two triangles. Classify the triangles by their angles <u>and</u> sides.

Triangle 1: (0, 0), (4, 0), (0, 4)

_____ and

_____

Triangle 2: (5, 5), (1, 8), (9, 4)

_____ and

_____

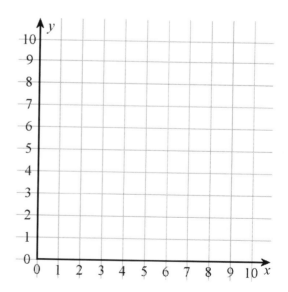

180

5. Fill in the missing parts in this tree diagram classification for triangles.

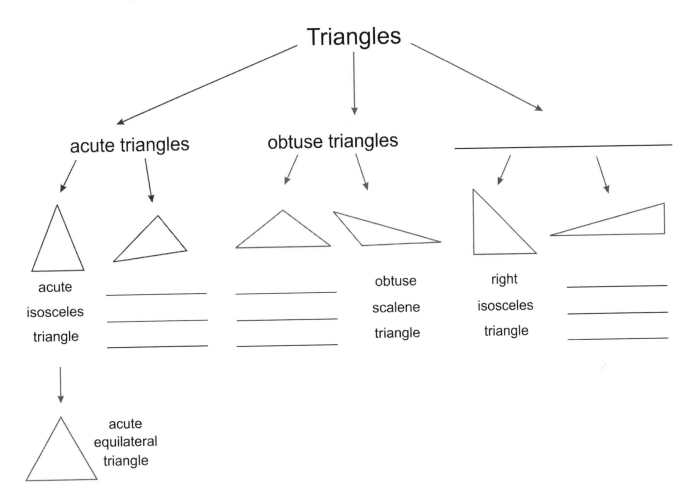

6. Sketch an example of the following shapes. You don't need to use a ruler.

    a. an obtuse isosceles triangle

    b. an obtuse scalene triangle

    c. a right scalene triangle

7. Plot in the grid

    a. a right isosceles triangle

    b. an acute isosceles triangle.

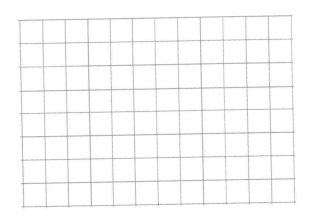

8. Make a guess about the angle measures in an equilateral triangle: _____°
   Measure to check.

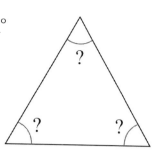

9. **a.** Could an equilateral triangle be a right triangle?
   If yes, sketch an example. If not, explain why not.

   b. Could a scalene triangle be obtuse?
   If yes, sketch an example. If not, explain why not.

   c. Could an acute triangle be scalene?
   If yes, sketch an example. If not, explain why not.

10. State whether or not it is possible to draw the following figures. (You don't have to draw any.)

   a. an obtuse equilateral triangle

   b. a right equilateral triangle

   c. an acute isosceles triangle

11. Measure all the angles in these isosceles triangles. Continue their sides, if necessary.
    Mark the angle measures near each angle.

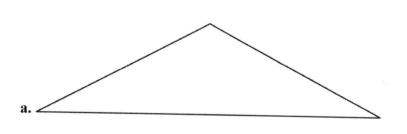

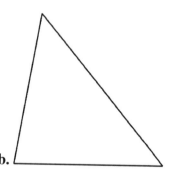

12. **a.** Draw any isosceles triangle.
    Hint: Draw any angle. Then, measure off the two congruent sides, making sure they have the same length. Then draw the last side.

   b. Measure the angles of your triangle. They measure _____°, _____°, and _____°.

13. Based on the last two exercises, can you notice something special about the angle measures?

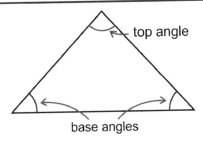

top angle

base angles

There are two angles in an isosceles triangle that have the SAME angle measure. They are called the **base angles.**

The remaining angle is called the **top angle**.

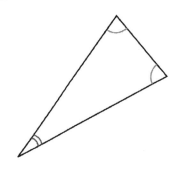

Can you find the top angle and the base angles in this isosceles triangle?

14. Draw an isosceles triangle with 75° base angles. (The length of the sides can be anything.)
Hint: Start by drawing the base side (of any length). Then, draw the 75° angles.

15. **a.** Draw an isosceles right triangle whose two sides measure 5 cm.
Hint: Draw a right angle first. Then, measure off the 5-cm sides. Then draw in the last side.

b. How long is the third side?

c. What is the measure of the base angles?

16. Draw a scalene obtuse triangle where one side is 3 cm and another is 7 cm.
Hint: Draw the 7-cm side first, then the 3-cm side forming any obtuse angle with the first side.

a. Draw two isosceles triangles with a 50° top angle. Your two triangles should not be identical.

Puzzle Corner

b. What is the angle measure of the base angles?

New Terms
• *equilateral triangle* • *isosceles triangle* • *scalene triangle*

Area and Perimeter Problems

Example 1. Find the area of the shaded figure.

The easiest way to do this is:
(1) Find the area of the larger outer rectangle.
(2) Find the area of the white inner rectangle.
(3) Subtract the two.

1. The area of the large rectangle is 7 cm × 10 cm = 70 cm².

2. We find the *sides* of the white rectangle by subtracting.

   The longer side of the white rectangle is
   10 cm − 5 cm − 1 cm = 4 cm.
   The shorter side is 7 cm − 2 cm − 2 cm = 3 cm.

   So, the area of the white rectangle is 4 cm × 3 cm = 12 cm².

3. Now we subtract to find the shaded area: 70 cm² − 12 cm² = 58 cm².

1. **a.** Find the area of the white rectangle.
   All lines meet at right angles.

   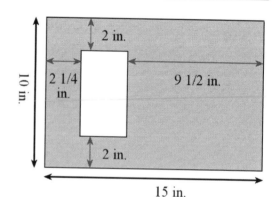

   b. Find the area of the shaded figure.

2. The image on the right shows a picture frame.
   Find the area of the actual frame (that is, of the shaded part).
   All lines meet at right angles.

   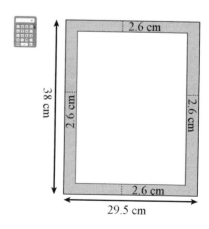

184

Example 2. Find the perimeter of the figure.

We need to find the length of *each* side and then add the
lengths. Start, for example, at the side marked with 1,
then go to the side marked with 2, then to side 3, and so
on, until you have "traveled" all the way around the figure.

Side 1 is 3 cm. Side 2 is 2 cm. Side 3 is 5 cm.
The total perimeter is:

3 cm + 2 cm + 5 cm + 5 cm + 4 cm + 1 cm + 4 cm + 4 cm = 28 cm.

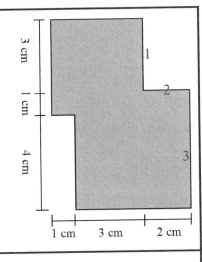

Example 3. Find the area of the figure.

Divide the figure into rectangles by drawing in it some additional lines.

Rectangle 1 has an area of 4 cm × 4 cm = 16 cm^2.

Rectangle 2 has an area of 3 cm × 4 cm = 12 cm^2.

Rectangle 3 has an area of 2 cm × 5 cm = 10 cm^2.

The total area is: 16 cm^2 + 12 cm^2 + 10 cm^2 = 38 cm^2.

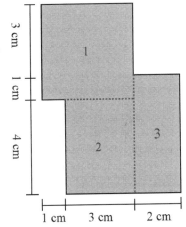

3. Find the area and the perimeter of this figure.
   All lines meet at right angles.

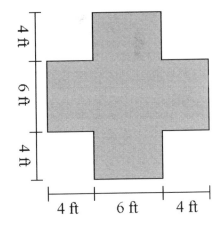

4. The perimeter of a rectangle is 42 cm.
   If the long side of the rectangle is 11 cm,
   how long is the shorter side?

185

5. Find the area and the perimeter of this figure.
   All lines meet at right angles.
   The dimensions are given in centimeters.

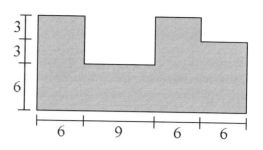

6. One side of a rectangular field measures 330 ft.
   A farmer fenced it with 910 ft of fencing.
   How long is the other side of the field?

7. The perimeter of a square is ½ mile.

   a. How long is one side of the square, in miles?
   Draw a sketch to help you.

   b. How long is one side of the square, in *feet*?

8. Find the area and the perimeter of this figure.
   All lines meet at right angles.
   The dimensions are given in inches.

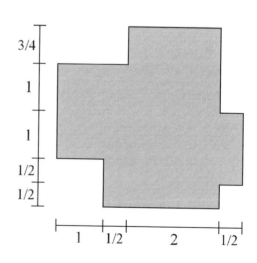

Volume

The **volume** of an object has to do with how much SPACE it takes up or occupies.

You have measured the volume of liquids using measuring cups that use ounces or milliliters. If we need to know the volume of a big object, such as a room, we cannot pour water into it to measure it with measuring cups. Instead, we use cube-shaped units or **cubic units**, and we simply check or calculate how many cubic units fit into the object.

 This little cube is **1 cubic unit.**

The volume of the figure on the right is six cubic units: V = 6 cubic units. Notice that one cube is not visible.

1. Find the volume of these figures in cubic units. "V" means volume.

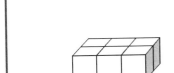

a. V = _____ cubic units

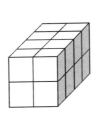

b. V = _____ cubic units

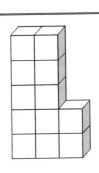

c. V = _____ cubic units

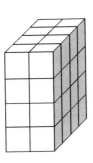

d. V = _____ cubic units

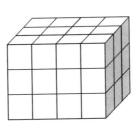

e. V = _____ cubic units

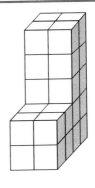

f. V = _____ cubic units

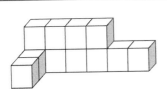

g. V = _____ cubic units

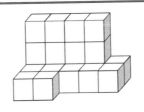

h. V = _____ cubic units

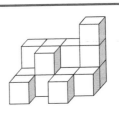

i. V = _____ cubic units

If each edge of this cube measures 1 cm, then the volume of the cube is **1 cubic centimeter**. This is abbreviated as **1 cm³**.

If each edge of the cube is 1 foot, its volume is **1 cubic foot.**

V = 1 cu. ft. or 1 ft³

If each edge of the cube is 1 inch, its volume is **1 cubic inch**.

V = 1 cu. in. = 1 in³

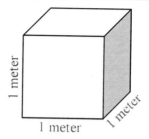

If each edge of the cube is 1 meter, its volume is **1 cubic meter.**

V = 1 m³

In general, if the edges of the cube are in certain units (such as inches, feet, centimeters, or meters), then the volume will be in corresponding *cubic* units.

If no unit is given for the edge lengths, we use the word "unit" for the lengths of the edges, and "cubic unit" for the volume. This "box" has a volume of 18 cubic units.

2. Find the total volume of each figure when the edge length of the little cube is given. Remember to include the unit!

The edge of each cube is 1 in.	The edge of each cube is 1 ft.	The edge of each cube is 1 cm.
	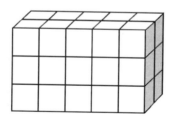	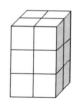
a. V = _____*3 in³*_____	**b.** V = _____	**c.** V = _____
The edge of each cube is 1 m.	The edge of each cube is 1 cm.	The edge of each cube is 1 in.
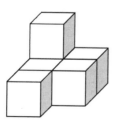	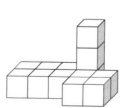	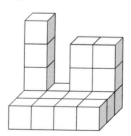
d. V = _____	**e.** V = _____	**f.** V = _____

This figure is called a **rectangular prism.** It is also called *a cuboid.*
It is simply a box with sides that meet at right angles.

Many people call the **three dimensions** that we measure "length,"
"width," and "height." Here we will use "width," "depth," and "height."

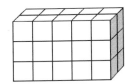

The **width** will be the dimension that runs left to right.
The **depth** will be the dimension that points away from you—into the paper, so to speak.
The **height** will be the dimension pointing "up" in the figure.

A way to find the volume of a rectangular prism by calculating

1) Can you figure out a way to find the number of cubes in
the *bottom* layer of this rectangular prism *without* counting?

You can multiply 5 × 2 = 10, which means multiplying
the *width* and the *depth*. The bottom layer has 10 cubic units.

2) After that, there is a way to easily find the *total* number of cubes in the rectangular prism
(its volume). Can you figure that out?

Take the number of cubes in the bottom layer, and **multiply that by how many layers there are**
(the *height*). There are 10 cubes in the bottom layer, and 3 layers. We get 10 × 3 = 30 cubic units.

3. Find the volume of these rectangular prisms by finding the amount of cubic units in the bottom layer
and multiplying that by the height (how many layers there are).

	a.	**b.**	**c.**	**d.**
Cubes in the bottom layer	8			
Height	4			
Volume	32			

4. If each little cube is 1 cubic inch, what is the total
volume of the outer box?

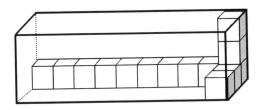

Notice what we did in these two steps:

(1) We multiplied the <u>width</u> and the <u>depth</u> to find the number of cubes in the bottom layer. **Multiplying the width and the depth** also gives us **the area of the bottom** (A_b)! For example, the bottom area of this cuboid is $4 \times 3 = 12$ square units.

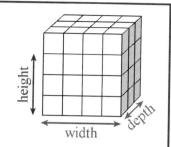

(2) We multiplied what we got from step 1 by <u>height</u>.

We ended up multiplying the bottom area by the height.
Or, looking at it in another way, we multiplied the width, the depth, and the height.

From that we get two **formulas** for the volume of a rectangular prism:

1. $V = w \times d \times h$ (volume is width × depth × height)

2. $V = A_b \times h$ (volume is area of the bottom × height)

5. Write the width, height, and depth of these rectangular prisms. Lastly, multiply those three dimensions to find the volume.

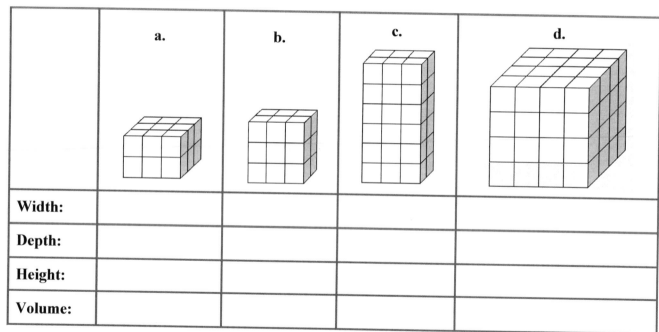

	a.	b.	c.	d.
Width:				
Depth:				
Height:				
Volume:				

6. Find the volume of the rectangular prisms above *if* their top layer was removed. Use cubic units. Use the formula $V = w \times d \times h$.

a. V = _____ × _____ × _____ = _____ cubic units

b. V = _____ × _____ × _____ = _____ cubic units

c. V = _____ × _____ × _____ = _____ cubic units

d. V = _____ × _____ × _____ = _____ cubic units

Now we can explain where the little raised "3" (the exponent) in cubic units comes from.

To find the volume of this cube, we multiply its width, the height, and the depth. This means multiplying 1 cm × 1 cm × 1 cm. Not only do we multiply number 1 by itself three times—we also multiply the *unit centimeter* (cm) by itself <u>three</u> times. The little "3" in cm^3 shows that.

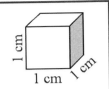

7. **a.** Sketch a rectangular prism with a volume of 4 × 2 × 6 cubic units.

b. Sketch a rectangular prism with a volume of 3 × 3 × 3 cubic units.

c. Sketch a rectangular prism with a volume of 2 × 5 × 4 cubic units.

8. Chris and Mia drew these rectangular prisms to match the expression 5 × 3 × 2. Who is right?

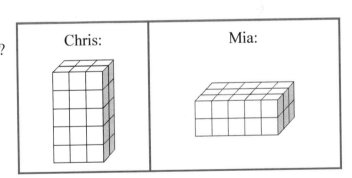

9. To calculate the volume of this kind of figure, think of it as consisting of *two* rectangular prisms. We calculate the volume of each separately, and then add. Which expression below matches the volume of this figure?

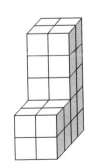

a. 2 × 3 × 2 + 2 × 2 × 3

b. 2 × 2 × 2 + 2 × 2 × 3

c. 2 × 2 × 2 + 2 × 2 × 5

Volume of Rectangular Prisms (Cuboids)

Study the two formulas for the volume of a rectangular prism:

1. $V = w \times d \times h$ (volume is width × depth × height)
   Some people use width, <u>length</u>, and height instead.

2. $V = A_b \times h$ (volume is area of the bottom × height)

The width, depth, and height need to be in the <u>same</u> kind of unit of length (such as meters). The volume will then be in corresponding cubic units (such as cubic meters).

Example 1. A room measures 12 ft by 8 ft, and it is 8 ft high. What is the volume of the room? What is the area of the room?

To find the area, we simply multiply the two given dimensions: A = 12 ft × 8 ft = 96 ft².
To find the volume, we can multiply the area by the height: V = 96 ft² × 8 ft = 768 ft³.

1. **a.** Find the volume of a box that is 2 inches high, 5 inches wide, and 7 inches deep. Include the units!

   V = ____5 in____ × _____ × _____ = _____

   b. Find the area and volume of a room that is 25 ft × 20 ft, and 9 feet high. Include the units!

   A = _____ × _____ = _____

   V = _____ × _____ × _____ = _____

2. Find the volume of a box that is

   a. 20 cm wide, 30 cm deep, and half a meter high.
   Note: you will need to convert the last measurement into centimeters before calculating the volume.

   b. 16 square inches on the bottom, and 6 inches tall.

3. The volume of this box is 30 cm³.
   What is its depth?

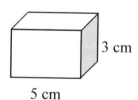

   3 cm

   5 cm

4. *Optional.* Measure the width, height, and depth of a dresser and/or a fridge. Find out its volume.

Volume is **additive**. What we mean by that is that we can ADD to find the total volume of a shape that is in several parts.

To find the total volume of the shape on the right, first find the volume of the top box, then the volume of the bottom box, and add the two volumes.

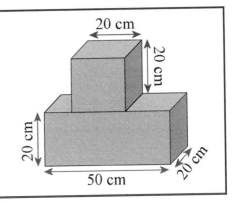

5. Find the total volume of the shape in the teaching box above.

6. This is a two-part kitchen cabinet. Its height is 2 ft and depth 1 ft. One part is 5 ft long, and the other is 4 ft long.

   a. Mark the given dimensions in the picture.

   b. Calculate the volume.

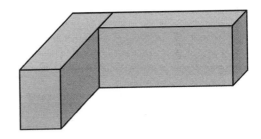

7. Design a box (give its width, height, and depth) with a volume of

   a. 64 cubic inches

   b. 1,200 cubic centimeters

8. The length and width of a rectangular box are 5 inches and 6 inches. Its volume is 180 cubic inches. How tall is it?

9. Find the volume of this building.

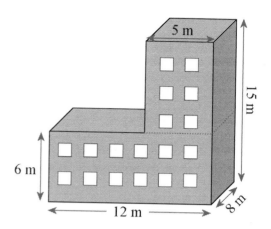

10. Find the total volume.

11. The picture shows an aquarium that is 1/4 filled with water.

   a. Find the total volume of the aquarium.

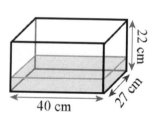

   b. Find the volume of the water in it.

12. John's room is 12 ft × 18 ft, and it is 9 ft high. The family plans to *lower* the ceiling by 1 foot.

   a. What will the volume of the room be after that?

   b. How much volume will the room lose?

13. **a.** Find the volume of this two-part bird cage.

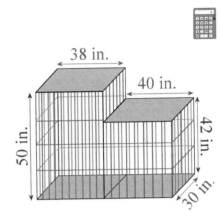

   b. One cubic foot is 1,728 cubic inches.
      Convert your answer from (a) into cubic feet, to three decimals.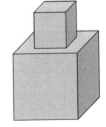

Puzzle Corner

The volume of the larger cube is 1,000 cubic inches.
The edge length of the smaller cube is half of
the edge length of the larger cube.

What is the combined volume of the two cubes?

A Little Bit of Problem Solving

1. Fill in the table, continuing the patterns.

   a. What is the area of the 11th square?

   b. What is the number of the square with an area of 10,000 m²?

Square number	Length of Side	Area
1	2 m	4 m²
2	4 m	
3	6 m	
4		
5		
6		

2. A wall is 16 feet wide and 10 feet high. It has one 3.5 ft × 4.5 ft window in it.

   a. Draw a sketch.

   b. Find the area of the window.

   c. What is the area of the actual wall (not including the window)?

   d. A gallon of paint covers 350 square feet of wall.
   How many whole gallons do you need to paint the wall?

   e. If instead of gallons, you buy paint by quarts, how many whole
   quarts of paint do you need to paint the wall?

3. This aquarium's dimensions are 4 ft × 3 ft × 3 ft.

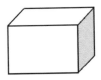

   a. How many cubic feet of water would fit in it?

   b. You fill it 8/9 full. How many cubic feet of water is that?

4. The volume of this cube is 8 cubic inches.
   How long is its edge?

5. What part of the whole design does the highlighted area make up?
   (Remember to simplify your fraction.)

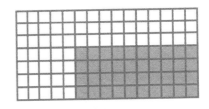

6. A certain room measures 15 ft × 20 ft. What part
   of the floor would a 5 × 4 ft carpet cover?

7. **a.** Draw a right triangle with 5 cm and 12 cm sides, using these steps:

    1. Draw a long line.
    2. Measure the 12-cm side on the line, marking it with two dots.
    3. Draw a perpendicular line through one of the dots. Use a proper tool!
    4. On the second line, measure and mark the 5-cm side.
    5. Draw in the third side of the triangle now.

   b. Find the perimeter of your triangle, in centimeters.

Mixed Review

1. Eric and Angela did some yard work together. They earned $80 and split it so that Eric got $12 more than Angela. How much did each one get?

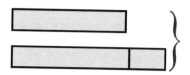

2. A bunch of five orchids costs $40 and a bunch of twenty daisies costs $40. Find the price difference between *one* orchid and *one* daisy.

3. Divide using long division. Check by multiplying.

a. $9{,}890 \div 46$	**b.** $71.5 \div 65$

4. Multiply and divide mentally.

a. $3 \times 0.25 = $ _____	**b.** $8 \times 0.08 = $ _____	**c.** $10 \times 0.009 = $ _____
d. $0.9 \times 8 = $ _____	**e.** $0.002 \times 5 = $ _____	**f.** $2 \times 0.3 \times 7 = $ _____
g. $0.8 \div 4 = $ _____	**h.** $100 \times 0.04 \times 2 = $ _____	**i.** $7.2 \div 8 = $ _____
j. $0.8 \div 0.4 = $ _____	**k.** $2 \div 0.01 = $ _____	**l.** $0.056 \div 7 = $ _____

5. Solve.

a. $6 \times \dfrac{1}{5} =$	**b.** $\dfrac{1}{3} \times \dfrac{2}{7} =$	**c.** $\dfrac{6}{11} \times \dfrac{1}{8} =$
d. $\dfrac{10}{15} \times \dfrac{5}{6} =$	**e.** $3 \div \dfrac{1}{5} =$	**f.** $5 \div \dfrac{1}{3} =$
g. $\dfrac{1}{5} \div 2 =$	**h.** $\dfrac{1}{10} \div 3 =$	**i.** $7 \div 5 =$
j. $4 \div 9 =$	**k.** $40 \div 3 =$	**l.** $62 \div 9 =$

6. A certain company makes skin salve in little jars that are
   1 3/8 inches tall. Those jars gets packed into boxes that
   are 6 inches tall. How many jars can be stacked on top
   of each other and still fit in the box?

7. Solve for *x*.

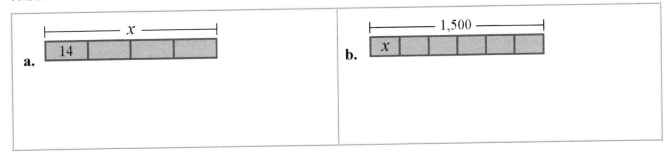

8. This morning one of her children is sick, so Mom
   is making only 2/3 of her usual recipe for pancakes.
   How much of each ingredient will she need?
   (*dl* stands for *deciliter*)

   What do you think she should do with the eggs?

   ┌─────────────────────────────┐
   │ Pancakes   │
   │   │
   │ 4 dl water   │
   │ 2 eggs   │
   │ 3 dl whole wheat flour   │
   │ (pinch of salt)   │
   │ 50 g butter for frying   │
   └─────────────────────────────┘

9. Split the pieces further, and cross out some pieces to show the subtractions. Solve.

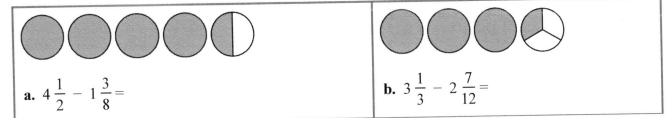

a. $4\dfrac{1}{2} - 1\dfrac{3}{8} =$

b. $3\dfrac{1}{3} - 2\dfrac{7}{12} =$

10. The ratio of green shirts to blue shirts is 3:5.

 a. What is the ratio of green shirts to all shirts?

 b. What is the ratio of blue shirts to all shirts?

 c. What is the ratio of blue shirts to green shirts?

11. The table shows how many adult and child visitors a small art museum had during one week.

 a. Calculate the total visitor counts.

 b. What was the difference in the total visitor count between the busiest day and the least busy day?

 c. Find the average number of adult visitors in a day. Give your answer to one decimal digit. Use your notebook for the long division.

 d. Find the average number of child visitors in a day. Give your answer to one decimal digit. Use your notebook for the long division.

Museum visitors			
Day	**Adults**	**Children**	*Total Visitors*
Monday	29	14	
Tuesday	23	10	
Wednesday	34	18	
Thursday	38	19	
Friday	35	19	
Saturday	57	25	
Sunday	63	31	
Totals			

 e. Make a double-bar graph of this data.

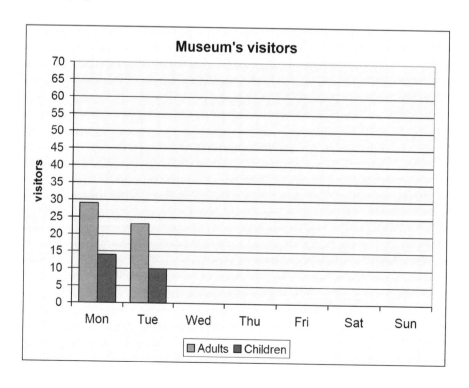

Review

1. Measure all the angles of the triangles. Then classify the triangles.

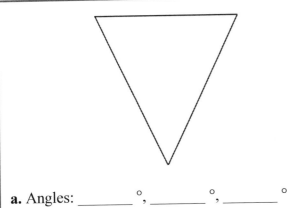 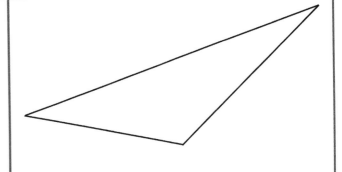

   a. Angles: _____ °, _____ °, _____ °

      Acute, obtuse, or right?

      Equilateral, isosceles, or scalene?

   b. Angles: _____ °, _____ °, _____ °

      Acute, obtuse, or right?

      Equilateral, isosceles, or scalene?

2. **a.** Draw an isosceles triangle with 50° base
   angles and a 7 cm base side (the side
   between the base angles).

   b. Measure the top angle.

      It is _____ ° .

   c. Find the perimeter of your triangle in
   millimeters.

3. Find the perimeter and area of this figure.
   All measurements are in inches.

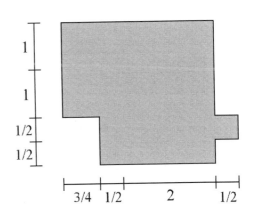

4. Name the different types of quadrilaterals.

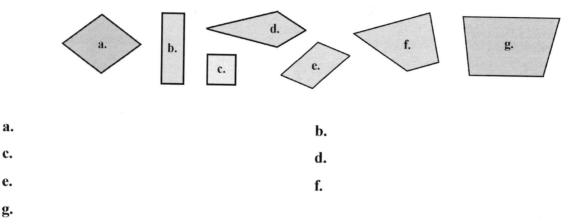

a.

b.

c.

d.

e.

f.

g.

5. Name the quadrilateral that...

   a. is a parallelogram and has four right angles.

   b. is a parallelogram and has four sides of the same length.

   c. has two parallel sides and two sides that are not parallel.

6. **a.** What is this shape called?

   b. Draw enough diagonals inside the shape to divide it into triangles.

   c. Number each of the triangles.

   d. Classify each triangle according to its sides (equilateral, isosceles, scalene) and according to its angles (acute, obtuse, right).

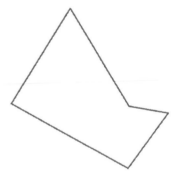

7. **a.** Draw an isosceles obtuse triangle.

   b. Draw a scalene acute triangle.

8. **a.** Draw a circle with its center at (2, 3) and
a radius of 2 units. Use a compass.

b. Draw another circle with its center at (6, 5)
and a *diameter* of 8 units.

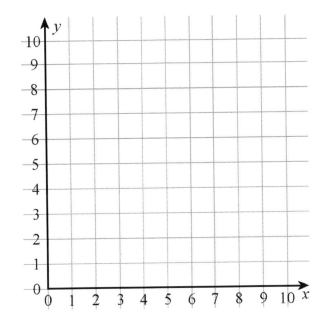

9. Find the volume of this rectangular prism, if…

a. …the edge of each little cube is 1 inch.

b. …the edge of each little cube is 2 inches.

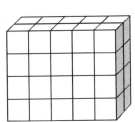

10. What is the height of this box, if its bottom dimensions are 2 cm × 4 cm
and its volume is 32 cubic centimeters?

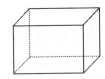

11. A gift box is 6 inches wide, 3 inches deep, and 2 inches tall.
How many of these boxes do you need to have a total volume of
108 cubic inches?

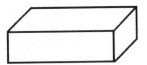

Puzzle Corner   The area of the bottom face of a cube is 16 cm². What is its volume?

203

Made in the USA
San Bernardino, CA
12 October 2018